Current Topics in Radiography Number 2

Edited by

Audrey Paterson FCR, MSc, TDCR, DMU
*Head of Department of Radiography,
Canterbury Christ Church College,
Canterbury, Kent, UK*

and

Richard Price FCR, MSc
*Head of Division of Radiography,
Division of Clinical Sciences,
University of Hertfordshire,
Hatfield, Hertfordshire, UK*

W.B. SAUNDERS COMPANY LTD
London · Philadelphia · Toronto
Sydney · Tokyo

W.B. Saunders Company Ltd 24–28 Oval Road
London NW1 7DX, England

The Curtis Center
Independence Square West
Philadelphia, PA 19106–3399, USA

55 Horner Avenue
Toronto, Ontario M8Z 4X6, Canada

Harcourt Brace
(Australia) Pty Ltd
30–52 Smidmore Street
Marrickville, NSW 2204, Australia

Harcourt Brace Japan Inc.
Ichibancho Central Building
22–1 Ichibancho
Chiyoda-ku, Tokyo 102, Japan

© 1996 W.B. Saunders Company Ltd

Chapter 3, Computed Tomography – A Spiralling Challenge in Radiological Protection, pp. 26–34, is copyright © 1996 of National Radiological Protection Board.

All rights reserved. No part of this publication may be reproduced, stored in a retrieval system or transmitted, in any form or by any means, electronic, mechanical, photocopying or otherwise, without the prior permission of W.B. Saunders Company Ltd, 24–28 Oval Road, London NW1 7DX, UK.

British Library Cataloguing in Publication Data is available

ISBN 0–7020–2202–0

This book is printed on acid-free paper

Typeset by Phoenix Photosetting, Chatham, Kent
Printed and bound in Great Britain by Biddles Ltd, Guildford and King's Lynn

Current Topics in Radiography
Number 2

616.0757/PAT

KERRY GENERAL
HOSPITAL
HEALTHCARE LIBRARY
06871842 16

CURRENT TOPICS IN RADIOGRAPHY

Editorial Board

Derek Adrian-Harris
Head of Centre for Radiography Education, University of Portsmouth, St Mary's Hospital, UK

Jane Bates
Superintendent Radiographer, Ultrasound Department, St James' University Hospital, Leeds, UK

Anthony R. Divers
Consultant Radiologist, St Albans and Hemel Hempstead NHS Trust, Hemel Hempstead, UK

Peter Hogg
Senior Lecturer, School of Radiography, University College Salford, UK

Nuala Martin
Business Manager, Radiology Department, Hammersmith Hospital, London, UK

Jeremy R. L. Nettle
Business & Development Manager, Salisbury Health Care NHS Trust, Salisbury District Hospital, UK

Elizabeth M. Warren
MRI Service Manager / Superintendent Radiographer, The MRI Centre, John Radcliffe Hospital, Oxford, UK

Whait
ational Manager, Centre for er Treatment, Mount Vernon ital, Northwood, UK

Alexander Yule
Department of Diagnostic Imaging, University Hospital of Wales, Cardiff, UK

Contents

Contributors		ix
Contents of previous volumes		xiii
Preface		xvii
1	Radiography: An Emerging Profession *Richard Price and Audrey Paterson*	1
2	Breast Screening – The Flawed Paradigm? *Derek Adrian-Harris*	14
3	Computed Tomography – A Spiralling Challenge in Radiological Protection *Paul C. Shrimpton*	26
4	Technology and Changing Practice in Radiotherapy *Kate Westbrook*	35
5	Future Developments in Magnetic Resonance Imaging *Stephen John Blackband*	48
6	Moral Dilemmas in Diagnosing Fetal Anomalies Using Ultrasound *Regina Fernando*	58
7	The Nature of Image Reporting *Philip J. A. Robinson*	70
8	Ethical Dilemmas in Radiographic Research *Val Challen*	83
9	The Patient's Charter and the Delivery of Quality Services *Marilyn Hammick*	96

10	Patients' Perceptions of Radiotherapy Treatment *Christine Mackenzie*	105
11	The Macmillan Radiographer in Cancer Care *Catherine Meredith and Lorraine Webster*	115
12	Diagnostic Imaging and Outreach Clinical Services *David Blenkinsop*	123
13	Professional Influences and Contracting for Quality of Care *Alexander Yule*	130
14	The Concept of Risk Management and its Applications in Radiography *Lynda Bennie*	139
15	Magnetic Resonance Imaging and Medico-legal Work in Cerebral Palsy *Catherine Westbrook*	148
16	Radiation Dose Limits – the UK Approach. Is it Sound? *Madge Heath*	164
17	Gamma Camera Imaging of Positron Emitting Isotopes *A. J. Britten and J. N. Gane*	174
18	MRI of the Breast *Joanna Linas and Elizabeth M. Warren*	196
19	Coronary Artery Imaging by Magnetic Resonance *Dudley Pennell*	228
20	The Role of Ultrasound in the Diagnosis and Management of Scoliosis *Alanah S. Kirby*	237
21	Telemedicine and Teleradiology in Primary Care: A GP's Perspective *John Wynn-Jones and Susan Groves-Phillips*	250
22	Current Trends in Nuclear Medicine *Daphne Glass and A. Michael Peters*	259

23	Managing Professional Development in Relation to Service Needs *Noelle Skivington*	270
24	Shared Learning: Improved Care and Professional Survival *Hazel Colyer*	279
25	Computed Radiography – Influences on Sensitivity and Latitude *Sue Farmer*	289

Authors' reply to editor's comment on chapter 15   299

Index   303

Contributors

Derek Adrian-Harris Head of Centre for Radiography Education, University of Portsmouth, St Mary's Hospital, Portsmouth PO3 6AD, UK.

Lynda Bennie Lecturer for Radiography, Department of PPR, Division of Radiography, Glasgow Caledonian University, Southbrae Campus, Southbrae Drive, Glasgow G13 1PP, UK.

Stephen John Blackband Senior Lecturer, Magnetic Resonance Imaging Centre, Hull Royal Infirmary, Anlaby Road, Hull HU3 2JZ, UK.

David Blenkinsop Radiology Service Manager, Essex Rivers Healthcare Trust, Colchester General Hospital, Colchester, Essex CO4 5JL, UK.

A. J. Britten Consultant Physicist, Department of Medical Physics, St George's Hospital, London SW17 0QT, UK.

Val Challen Senior Lecturer, Department of Radiography and Imaging Sciences, University College of Saint Martin, Bowerham Road, Lancaster LA1 3JD, UK.

Hazel Colyer Programme Director MSc Interprofessional Health and Community Studies, Canterbury Christ Church College, North Holmes Road, Canterbury, Kent CT1 1QU, UK.

Sue Farmer Clinical Lecturer, Division of Radiography, University of Hertfordshire, Hatfield Campus, College Lane, Hatfield, Herts AL10 9AB, UK.

Regina Fernando Senior Lecturer, Division of Radiography, School of Health and Human Sciences, University of Hertfordshire, Hatfield Campus, College Lane, Hatfield, Hertfordshire AL10 9AB, UK.

J. N. Gane Department of Medical Physics, St George's Hospital, London SW17 0QT, UK.

Daphne Glass Department of Radiology, Hammersmith Hospital, Du Cane Road, London W12 0HS, UK.

Susan Groves-Phillips Research Associate, Centre for Health Informatics, University of Wales, Aberystwyth, Dyfed SY23 3AS, UK.

Marilyn Hammick Principal Lecturer, School of Radiography, Joint Faculty of Healthcare Sciences, Kingston University and St George's Hospital Medical School, Penrhyn Road, Kingston upon Thames, Surrey KT1 2EE, UK.

Madge Heath Senior Lecturer, Centre for Radiography Education, University of Portsmouth, St Mary's Hospital, Milton Road, Portsmouth PO3 6AD, UK.

Alanah S. Kirby Senior Research Radiographer, Department of Human Morphology, Medical School, Queen's Medical Centre, Nottingham NG7 2UH, UK.

Joanna Linas Deputy Superintendent Radiographer, MRI Centre, John Radcliffe Hospital, Headington, Oxford OX3 9DU, UK.

Christine Mackenzie Research Fellow, Health Research Centre, Faculty of Social Science and Education, Middlesex University, Queensway, Enfield, Middlesex EN3 4SF, UK.

Catherine Meredith Department of Physiotherapy, Podiatry and Radiography, Glasgow Caledonian University, Southbrae Campus, Southbrae Drive, Glasgow G13 1PP, UK.

Audrey Paterson Head of Department of Radiography, Canterbury Christ Church College, Canterbury, Kent CT1 1QU, UK.

Dudley Pennell Senior Lecturer and Honorary Consultant, MRI Unit, Royal Brompton Hospital, Sydney Street, London SW3 6NP, UK.

A. Michael Peters Department of Diagnostic Imaging, Hammersmith Hospital, Du Cane Road, London W12 0HS, UK.

Richard Price Head of Division of Radiography, School of Health and Human Sciences, University of Hertfordshire, College Lane, Hatfield, Herts AL10 9AB, UK.

Philip J. A. Robinson Consultant Radiologist, St James's University Hospital, Leeds LS9 7TF, UK.

Paul C. Shrimpton Principal Scientific Officer, Medical Dosimetry Group, National Radiological Protection Board, Chilton, Didcot, Oxon OX11 0RQ, UK.

Noelle Skivington Radiology Services Manager, The Royal Free Hospital, Pond Street, London NW3, UK.

Elizabeth M. Warren MRI Service Manager/Superintendent Radiographer, MRI Centre, John Radcliffe Hospital, Headington, Oxford OX3 9DU, UK.

Lorraine Webster Macmillan Radiographer, The Beatson Oncology Centre, Glasgow, UK.

Catherine Westbrook Imaging Research Assistant, Education and Research Co-ordinator in Magnetic Resonance Imaging, University Department of Radiology, John Radcliffe Hospital, Oxford OX3 9DU, UK.

Kate Westbrook Radiotherapy Services Manager, Bristol Oncology Centre, Horfield Road, Bristol BS2 8ED, UK.

John Wynn-Jones General Practitioner, CME Tutor, Montgomery Medical Centre, Well Street, Montgomery, Powys SY15 6PF, UK.

Alexander Yule Radiology Directorate Manager, University Hospital of Wales Healthcare NHS Trust, Heath Park, Cardiff CF4 4XW, UK.

Contents of previous volumes
No 1

Where Are We Now?
Olive M. Deaville

The Future of Ionizing Radiation Medicine in Diagnosis and Therapy
Elizabeth J. McLean

Are Radiologists Really Necessary?
Anthony R. Divers

Changing Practice in Radiography
Lorraine Nuttall

Patient Focus in Radiography – the Challenges
Diane Carney

Radiographic Reporting in Diagnostic Imaging
Lyn McKay

Developing Quality Standards in Radiography
David Dewitt

Quality Accreditation – What Cost?
Stephen A. Evans

Dose Reduction in Diagnostic Imaging
Timothy Reynolds

Audit of Professional Practice
Neil Prime

Digital Imaging – State of the Art and Future Potential
Andrew Todd-Pokropek

Computed Radiography – the Conquest Experience
David Baker

Image Transmission and Picture Archiving and Communication Systems – the Potential
Nuala Martin and Anne Hemingway

Trauma Imaging
Miles J. Woodford

The Value of the 18-week Scan in the Second Trimester of Pregnancy
Josephine Swallow

Intraoperative Ultrasound in Hepatic Resection
Jane A. Bates and Rose Marie Conlon

Magnetic Resonance Imaging Investigation of Arachnoiditis
Elizabeth M. Warren

Pseudo-dynamic Magnetic Resonance Imaging in Orthopaedics
Susan G. Moore

Spiral Computed Tomography: Current Applications and Future Potential
Leonie S. Paskin

Perfusion Computed Tomography
Vivian Wood

Positron Emission Tomography – a Clinical Tool?
John Lowe

The Future of Antibody Imaging, Research and Clinical Applications
Keith E. Britton and Maria Granowska

Conformal Radiotherapy
Lynne Sterry

Continuous Hyperfractionated Accelerated Radiotherapy
Cathy Williams

A Multidisciplinary Approach to Cancer Patients
Judy Young

Imaging Services in the Community
Alexander J. M. Cavenagh

Preface

Introducing the first edition of *Current Topics in Radiography* last year, we were conscious of how much its success depended upon our contributors and our readers; our contributors for their expertise and their uncompromising views and our readers for taking an interest in those views. We were very pleased, therefore, that so many of you found the book interesting and useful. We were also very grateful to those of you who wrote or telephoned to discuss the first volume. We have taken on board your comments and these have helped us in developing the conceptual framework of *Current Topics in Radiography*. W.B. Saunders, the publishers, were also very pleased with Volume 1; hence Volume 2. We trust you will find this volume as stimulating as the first.

As with Volume 1, we are indebted to our Editorial Board and to the authors who have given much of their time to produce what we believe is a worthy successor. For this volume, the Editorial Board remains unchanged. However, two members of the board retired during the year because of changes in employment. We would like to take this opportunity to give special thanks to Stephen Evans and to Jeremy Nettle and to wish them well for the future.

The theme of many of the chapters in the first volume reflected the developing roles of radiographers in all areas of practice, as well as the impact of changing technology on those roles. Twelve months on, these issues remain at the top of the agenda. The current volume seeks to advance the debates about changing roles and changing technologies by reflecting, yet further, on the practice of radiographers and their interaction with other professional groups and with patients. Across the spectrum of changing practice, many questions that relate to ethical and professional matters are emerging. *Current Topics in Radiography* No. 2 has attempted to raise some of these and we trust we have been successful in doing so.

Thanks are due to the considerable efforts of both our Editorial Board and, especially, our authors. We believe this issue has captured, at least in part, the mood of the profession and has identified some of the issues and concerns that are uppermost in the minds of practitioners. There are an increased number of editorial comments attached to the chapters compared with Volume 1. Our aim is to assist in continuing and developing the debates raised within those chapters.

Last year was a year of celebration marking the first century of X-radiation in diagnosis and treatment following Röntgen's famous discovery. X-rays and radium were the tools of radiographers in the early part of the century. Yet, by its close, complex linear accelerators dominate the practice of therapeutic radiographers while the spectrum of diagnostic radiographers' practice encompasses nuclear medicine, ultrasound and magnetic resonance imaging. The changes that occurred during that first century were staggering but even those impressive achievements may be small in comparison with what lies ahead in the next 100 years. In the immediate future, boundaries between diagnosis and treatment will blur considerably and the use of non-ionizing radiation techniques for imaging, intervention and treatment will develop further.

While the changes presaged should result in simpler, safer and more accurate care for patients, the demands on practitioners will be more sophisticated and complex. Inevitably, radiographers will need to establish new roles and accept new and more responsibilities. As in the first century, some of these new roles and responsibilities will be contentious and contested. *Current Topics in Radiography* attempts to address some of these issues and our hopes are that it will provide a focus for the continuing debate.

Finally, we would like to thank our readers. Ultimately, it is only you who will determine whether *Current Topics in Radiography* meets your needs. We hope you will enjoy this volume and will find it exciting and stimulating. Please let us know what you feel about it and please think about making your views known through its pages.

Audrey M. Paterson
Richard Price

1
Radiography: An Emerging Profession

Richard Price and Audrey Paterson

INTRODUCTION

In the early years of radiation medicine there was no clear distinction between radiography and radiology and the terms were used interchangeably, at least, until the 1930s (Larkin, 1983). However, within ten years of Röntgen's discovery moves were made to establish boundaries between the medical and non-medical practitioners (Arthur and Muir, 1909) and in 1925, not long after the establishment of the Society of Radiographers, these moves were realized (Moodie, 1970). Subsequently, radiography was reduced to mean the production of radiographs and was practised by non-medically qualified, or technical personnel; radiographers working at the behest of medical practitioners. Radiology came to mean the medical interpretation of radiographs and became the exclusive domain of medically qualified staff who began to develop the specialty of radiology.

Early developments in x-ray work have been documented by Burrows (1986), although he did not delve into the struggle for professional domination amongst the various occupational groups that provided those early x-ray services. Larkin (1978, 1983) analysed the development and social organization of radiography from an occupational perspective and illustrated the complete control that came to be exerted by the medical profession.

This chapter will seek to examine the status of radiography

throughout its history and evaluate the professional status of radiography at the end of the twentieth century.

MEDICAL DOMINANCE

The early non-medical x-ray workers, or radiographers, were responsible for the care of their patients, the operation of the x-ray equipment and, significantly, for the reporting of radiographs. This recognized implicitly their specialized knowledge and skills, as well as their responsibility and accountability to patients and to the wider society in which they practised. These were powerful indicators of autonomy and established the first radiographers as practitioners in their own right and radiography as a profession. This was not to last beyond the 1920s and radiographers were, effectively, de-skilled and de-professionalized as a result of the medical dominance arising from the decision forced upon the Society of Radiographers in 1925 to prohibit its non-medical members from reporting upon their own radiographs.

The 1930s brought about the Board of Registration of Medical Auxiliaries (BRMA), the precursor of state registration. As Larkin (1983) pointed out, the main concern of the Society of Radiographers was to maintain parity of subordination with other paramedical organizations rather than to avoid the increasing, and possibly malign, control of the British Medical Association (BMA) which it exercised through the BRMA. According to Larkin, radiologists were beginning to grow in strength and probably brought pressure to bear on the BMA so that it passed a resolution to restrict the interpretation of radiographs to radiologists or properly qualified general practitioners only.

Larkin believed that the entrance of the Society of Radiographers to the BRMA added the final dimension to medical dominance of radiography in the inter-war period. The Articles of Association of the BRMA stated that if the majority of members of the Council of the Society of Radiographers were not registered medical practitioners then the BMA shall appoint as many practitioners as necessary to ensure a majority. Medical membership of the Council of the Society of Radiographers was able to ensure through examination systems and education committees that the medical education of radiographers was limited to developing an expertise in the mechanics of production. As Larkin (1978) concluded:

The refusal to generalize education was clearly restrictive in intent.

The definition of medical and non-medical roles within radiation medicine; medical membership of the Council of the Society of Radiographers, and the integration of the Society into the BRMA closely supervized by the BMA were the delineating factors in the, almost deliberate, de-skilling and de-professionalizing of radiography. One further point of relevance was the question of the purpose of the Society of Radiographers. Was it established as a professional or governing body or as a means of controlling non-medical practitioners in radiography? Whatever the answer, the right to claim professional status had to be earned and, clearly, radiography had lost this right by the early 1930s.

DICHOTOMY OF INTERESTS

Medical domination of radiography was evident not only in the field of education but also in relation to the pay and conditions of employment of radiographers. In the 1930s attempts to improve conditions of employment and, hence, the professional status of radiographers were not helped by the attitude of the Society of Radiographers. Considerable and collective pressure for change was exerted by the rank and file members but at a special meeting of the Society, the President, Dr Leo Rowden, a radiologist, stated that radiographers required the goodwill of radiologists and nothing should be done to affect this. The President's view was supported by some radiographer members of the Council and it was stated that the salary of the radiographer was to depend upon the efficiency of the individual. Presumably, this efficiency was to be judged by the radiologists and was to be dependent upon retaining their goodwill.

It was Larkin's view that very prominent officials discouraged radiographers from taking the kind of collective action on pay and conditions of employment that was more characteristic of the BMA and specialist medical groups. It appeared that while it was possible for there to be a convergence of occupational and professional interests for the medical professions this was not permissible for radiographers. Sadly, the Society of Radiographers conformed to this view.

Here is a clear dichotomy of interest between professional and trade, or employment, interests. The professional ethos must be concerned with service to, and non-exploitation of, the population served by the particular professional or occupational group rather than the professional status of the group as might be evidenced by its pay and conditions of employment. For much of the 1920s and 1930s and indeed, later, professional status as evidenced by pay and conditions of employment was beyond the radiographers' control and low pay with poor employment conditions perpetuated. The links between the proper professional aspirations of public service and beneficence, and terms and conditions of employment do not appear to have been explored. But, poor conditions of employment and salary structures are unlikely to motivate radiographers to pursue their occupation at the highest level. Radiographers, it would seem were, for many years, little more than an organized occupational group maintained and controlled by their medical benefactors.

RE-ESTABLISHMENT OF RADIOGRAPHY AS A PROFESSION

The move to re-establish radiography as a profession in its own right and to legitimize and secure the professional aspirations of radiographers was slow but a breakthrough came in 1960 with the Professions Supplementary to Medicine Act (PSM) when radiography was recognized by statute. Moodie (1970) claims that this Act gave radiography full professional status within the National Health Service (NHS) and commented that it 'was a remarkable achievement and on paper at least it put an end to the master servant relationship with the doctor.' In reality, little changed and it is arguable whether radiography in 1960 displayed the characteristics normally ascribed to professions.

CHARACTERISTICS OF PROFESSIONS

Many authors, including Hugman (1991) have postulated characteristics and definitions of professions and much debate has taken place about their appropriateness. Indeed, Illich (1976) goes further and questions the appropriateness of the existence of professions, particularly in the field of health. One approach to defining professions, was that taken by Millerson (1964) who suggested that professions exhibit five traits. These were:

- a body of knowledge;
- monopoly;
- testing of competence;
- code of conduct;
- ideology of service.

Other definitions include those of Wilensky (1964) who argued the functionalist view in which the emergence of professions is a dynamic, five stage process passing from the emergence of an occupational group to the situation whereby the group achieves recognition and protection through engaging in political activity; Larson (1977) described three dimensions of professions, those being the cognitive (knowledge and skills), the normative (services and ethics) and the evaluative (autonomy and prestige) dimensions; and Blane (1986) with his continuum approach where the so-called ideal professions of medicine and law occupy one extreme of the continuum while the occupations of, for example, plumbing and bricklaying occupy the other. Superficially at least, these definitions may be used to characterize radiography. But does superficial accord with some, or all, of the traditional characteristics of professions justify labelling radiography a profession and, perhaps more importantly, its practitioners as professionals?

Hickson and Thomas (1969) did not try to answer the question of what constitutes a profession, arguing that the question was sterile. Of much greater value, they believed, was to consider how professional was a particular organization relative to key identifiable characteristics. They attempted to measure professionalization in Britain by reviewing professional characteristics of 43 qualifying associations, including radiography, using a 13 point, professional characteristics scale. In 1969, radiography ranked 35th of the 43 organizations assessed, scoring 2 on the professional characteristics scale used (range 0–13, mean 5.9). Scores achieved by other qualifying associations in the health arena included a score of 13 by the Royal College of Obstetricians and Gynaecologists; 13 by the Royal College of Surgeons; 11 by the then College of General Practitioners; 7 by the Chartered Society of Physiotherapists, and 5 by the Society of Chiropodists.

The study also found a positive correlation between the degree of professionalization and the age of an association. At the time of the study the Society of Radiographers had been in existence for 47 years while the organizations surveyed ranged in age from 15 years for the College of General Practitioners to 667 years for the Inns of Court. It is interesting to note that while the Chartered Society of Physiotherapy and the Society of Chiropodists scores were considerably higher than the score for the Society of Radiographers, they had been in existence for similar periods of time. Such a comparison supports the view that the professional status conferred on radiography by the PSM Act 1960 was, at most, tenuous.

Professional status and the recognition of that status has been an important issue for many workers who espouse to be professionals and radiographers have been no exception. But what gives an occupational group the right to be called a profession? What conditions must prevail? In trying to answer these questions a definition of profession was presented in *Radiography News* by Aldridge (1986) in order to promote the image of radiography:

> *A field of endeavour or set of activities based on a body of theoretical understanding and empirical research intended to provide service to a set of people called clients from which practitioners may receive remuneration. It includes a set of standards that define the capacity to practice professionally, to guarantee clients a minimum level of service and to provide a basis on which members of the profession may be censured for harmful exercise of their skills and misuse of their knowledge.*
>
> *It is directly involved in relating theory and practice and includes expectations that member are responsible for regularly updating their skills and knowledge and that education and training programmes are to be regularly exercised and revised on the basis of advance in theory and practice.*

More generally, Blane (1991) has proposed conditions that professions have in common (see Table 1.1). He also claimed that professions are slow to change and tend to be the focus of their members' self identity and loyalty.

RADIOGRAPHY: A PROFESSION?

It is pertinent to consider how radiography measures up to Blane's conditions.

Table 1.1 Conditions that professions have in Common (Blane, 1991)

1. Professions are found in the highly skilled sector of the labour market. They possess a body of knowledge to which they add by research and which is passed on to trainees in institutions controlled by the profession usually in a university setting.
2. They usually have a monopoly of their field of work which depends upon state registration.
3. Professions have considerable autonomy in organizing, defining the nature of, and developing their work: a freedom from outside control which is defended on the grounds that only a member of the same profession is competent to assess a professional's work.
4. Professions espouse a code of ethics which prohibits the exploitation of clients and regulates intra-professional relations.

His first condition contains four characteristics. These are the sector of the labour market occupied; a body of knowledge; the development of knowledge by research; and the control of the education of trainees, usually in a university setting.

Radiography is found in the highly skilled sector of the labour market and it does possess a specific body of knowledge. The promotion of a research ethos has always been an object of the Society, and later the College of Radiographers but is only just beginning to influence the professional culture and, hence, the knowledge base of the profession. Real engagement with research at practitioner level has only begun to emerge in the past ten years and only significantly in the past five years. Significant control over the education of radiographers is exerted through the Joint Validation Committee of the College of Radiographers and the Radiographers Board of the Council for the Professions Supplementary to Medicine (CPSM). This control of pre-registration education has weakened since the transfer to the higher education sector but most would argue that it was a worthwhile price to pay. Overall, it would appear that radiography begins to meet the characteristics within Blane's first condition.

Blane's second condition of monopoly based upon state registration has been satisfied since the PSM Act of 1960. Radiography should not, however, be complacent. Since the advent of the Act, practice has changed considerably and other occupational groups infringe the monopoly radiographers once exercised, at least, in the broad field of diagnostic radiography. For example, nuclear medicine technologists using ionizing radiations work in nuclear medicine departments, sometimes with radiographers but not infrequently independently. Additionally, a wide range of occupational groups, including midwives, cardiac technicians and vascular technologists, can be found carrying out ultrasound imaging examinations (Scott-Angell and Chalmers, 1992) so impinging on the practice of radiographers.

The extent to which the third condition is met is debatable. Radiographers, traditionally, have worked closely with radiologists and under their control, which resulted in the loss of professional status during the 1920s and 1930s. Radiographers have, of course, always been responsible for controlling work in a particular x-ray, imaging or treatment room and for the selection of a particular technique. However, they have not always determined the radiographic projections that should be taken and have usually been obliged to follow protocols which have been drawn up by medical staff, often without the central involvement of radiographers. Even with the newer imaging modalities of computed tomography (CT) and magnetic resonance imaging (MRI) radiographers often follow protocols determined by medical staff. Radiography is, however, changing and traditional work practices are being challenged. For example, sonographers now conduct their own clinical sessions, as do radiographers who undertake barium enema sessions (Paterson, 1995). These changes suggest that radiography is beginning to exercise autonomy in organizing, defining and developing its practices.

Radiography has permitted its work to be assessed by another profession. This has occurred every time that other profession (radiology) has reported on radiographs produced by radiographers, currently the standard working practice in radiography. This process is repeated when other medical staff interpret the radiological opinions given, so there is no clear basis on which there is, or should be, complete autonomy for any profession in terms of reporting. However, the issue of radiographers producing reports on their own work is a crucial one and will be a major determinant of the autonomy of the profession of radiography.

Blane's fourth set of characteristics suggests that professions espouse ethical codes of conduct and practice. Both the professional body, the College of Radiographers, and the statutory body, the Radiographers Board of the Council for Professions to Medicine, issue codes to govern the practice of radiographers; the Code of Professional Conduct (1994) and the Statement of Conduct (1995) respectively. Certainly, these prohibit expressly the exploitation of radiographers' patients and, further, require that radiographers' practice is to the benefit of patients. Intra-professional relations are also controlled; so it would seem, on first examination, that radiography meets this fourth set of conditions.

The claim that professions are slow to change is exemplified in radiography. It is almost forty years since the Professions Supplementary to Medicine Act of 1960 was first mooted, yet it is only now that radiography (in its post 1925 era) is beginning to match the conditions that characterize a profession, as identified by Blane (1991). It is clear that radiography must defend and consolidate its emerging professional status more vigorously in the future than it did in the past if it is to meet successfully the escalating technological and employment challenges which, undoubtedly, lie ahead.

OBLIGATIONS OF PROFESSIONS TO THE PUBLIC

All characterizations and definitions of professions suggest an ideology of public service. These were expounded by Lord Benson in a debate in the House of Lords in 1992. Benson identified nine obligations of a profession to the public (Table 1.2).

Benson does not attempt to define professions but, like Hickson and Thomas, he proposed a series of professional characteristics, which he refers to as obligations to the general public.

Of the nine obligations only the eighth refers directly to the members of a profession and it is debatable whether the eighth obligation is met by the majority of radiographers. How many radiographers exhibit in their practice independence of thought and outlook, and speak their minds without fear or favour? How many, in relation to their work, allow others to control and dominate that work inappropriately? It must be said, however, that it is far easier for professional bodies, including the College of Radiographers, to claim to fulfil Benson's

Table 1.2 Obligations to the public (Benson, 1992)

1. The profession must be controlled by a governing body which in professional matters directs the behaviour of its members.
2. The governing body must set adequate standards of education as a condition of entry and thereafter ensure that students obtain an acceptable standard of professional competence. Training and education is continued throughout the members' professional life.
3. The governing body must set the ethical rules and professional standards which are to be observed by the members. These should be higher than those established by the general law.
4. The rules and standards enforced by the governing body should be designed for the benefit of the public and not for the private advantage of the members.
5. The governing body must take disciplinary action including, if necessary, expulsion from membership should the rules and standards it lays down not be observed or should a member be guilty of bad professional work.
6. Work is often reserved to a profession by statute – not because it was for the advantage of members but because, for the protection of the public, it should be carried out only by persons with the requisite training, standards and disciplines.
7. The governing body must satisfy itself that there is fair and open competition in the practice of the profession so that the public are not at risk of being exploited.
8. The members of the profession, whether in practice or in employment, must be independent in thought and outlook. They must be willing to speak their minds without fear or favour. They must not allow themselves to be put under the control or dominance of any person or organization that could impair that independence.
9. In its specific field of learning a profession must give leadership to the public it serves.

professional obligations than for individuals to match on a daily basis the criteria set out in obligation 8. Inevitably, ruling bodies function a step away from practice which provides a buffer not available to individuals faced with difficult decisions and conditions under the pressures of everyday working. Nevertheless, the actions and status of a profession are legitimized by the body corporate and, particularly, by its leading practitioners.

Obligations, 1 to 5 and 7 refer to the governing body of a profession. For radiography, the general assumption is made that the Society and College of Radiographers fulfil the role of governing bodies but obligation 5, in particular, introduces a dilemma for the Society and College and raises legitimate concerns about their competency to undertake governing roles. The Society of Radiographers is a registered trade union while the College of Radiographers is a registered charity for educational and professional purposes. To complicate matters, the College does not, itself, have members and is wholly owned by the Society. The College has its Code of Professional Conduct (1994) which promotes standards of professional behaviour and is for the benefit of patients and the general public, yet the Society of Radiographers in its trade union capacity is obliged to defend members accused of misconduct. Conflict between the functions of the Society and the College is inevitable here, and is exacerbated by the fact that the ruling councils of the Society and the College comprise the same individuals. The critical test of competency would arise in a case where allegations of misconduct were made and found against a member. The Society, in its trade union capacity, is obliged to defend the member but, equally, the professional body (the College) must investigate the allegations and consider whether the individual is fit to remain a member of the profession. Potentially, at least, a very serious conflict of interest arises.

Similarly, where misconduct against a member of the Society of Radiographers is alleged, it is likely that the individual would be reported to the statutory body, the Radiographers Board, which must, in such circumstances, carry out a full investigation. Should the radiographer accused be found guilty of infamous conduct, their name may be removed from the State Register. It is likely that the Society of Radiographers would defend its member at any hearing but what is the role of the Society in this case? Does it mount an appeal against the decision to remove the radiographer from the register, or does it change its position relative to supporting the member and seek to remove the guilty party from membership as is implied by Article 7. (2) (Society of Radiographers, 1990)? What is the role of the College of Radiographers in such circumstances? There must be no argument or room for doubt. The Society and the College of Radiographers must act according to their Articles of Association and remove offending members from membership. Failure to act in the interest of patients and the general public would be a derogation of responsibility; would invalidate the code of professional conduct, and would nullify all claims to be the governing and professional body for radiography.

Lord Benson makes it clear that for a professional body to be self-regulating it must observe all nine obligations, yet radiographers cannot join their professional body, the College of Radiographers, directly. Instead, they must join the Society of Radiographers which, as trade union, exists primarily to serve its members. If radiography is a profession its prime obligation must be to the public it serves and it must have a governing body that recognizes this unambiguously. The question remains, therefore, – what, or who, comprises the governing body of radiography? It cannot be the statutory body, the Radiographers Board of the CPSM, which has registrants rather than members, and which operates only within the narrow and restrictive confines of the Professions Supplementary to Medicine Act, 1960 and by limited precedent. Statutory bodies rarely provide the necessary professional leadership and, frequently, portray the characteristics of quangos. The governing body cannot be the Society of Radiographers with its clear trade union obligations to its members; and it is difficult to see how it can be the College of Radiographers while its relationship with the Society of Radiographers remains so intimate.

Urgency is attached to the question posed above when it is realized that the College of Radiographers is likely to introduce a comprehensive system of continuing professional development during 1996, so enabling Benson's second obligation to be met fully. Inevitably, systems of continuing professional development infer penalties for those members of a profession who do not engage in the required lifelong learning or who fail to meet the designated standards. It is difficult to see how the College of Radiographers will be able to ensure rigour within its proposed continuing professional development strategy without distancing itself overtly from the Society of Radiographers.

CONCLUSION

The dichotomy between trade union and professional aspirations in radiography first raised itself in the 1930s and has remained ever since. Can a professional body have this duality of interest? Neither Blane nor Benson comment or provide insight. The Royal College of Nursing (RCN) has recently given up its pledge never to strike. Does this de-professionalize nurses? Or is this duality of purpose, of both trade and professional aspirations, consistent with a modern understanding of the nature of professions? The case may be made that organizations such as the Society and College of Radiographers and the Royal College of Nursing can only protect their patients against what they see as the injustices of government and employers by ensuring that the salaries and conditions of employment of their members are adequate to guarantee a high level of public service.

Perhaps, it is timely to review the work of Hickson and Thomas and to take into account the refinements to professional characteristics and obligations introduced more recently by Blane and by Benson. A suitably updated repetition of the 1969 survey of professions conducted by Hickson and Thomas may well reveal that those professions that ranked highest in 1969 have retained not dissimilar scores while radiography has improved its score significantly. Even so, it is unlikely that radiography will have reached the position on the continuum of professions occupied by the medical and legal professions. Of course, it may not wish to if its principal obligation is to remain one of public service.

Notwithstanding the relationship between the Society of Radiographers, the College of Radiographers and the Radiographers Board, radiography is now in its strongest position yet to claim professional status. However, there remain matters to be resolved and, in particular, vision is required to clarify the governance of the profession so that its professional status cannot be challenged in the future and to ensure that its obligations to the public can be upheld unequivocally. The question for radiography to address, therefore, is no longer whether it is a profession but, rather, what comprises the professional body and who comprises the governing council? The obligations which apply to a profession could be said to be met by the College of Radiographers but it remains an enigma while it remains subsidiary to the Society of Radiographers.

References

Aldridge, C.A. (1986) The professional image. *Radiography News*, October, 10.
Arthur, D., Muir, J. (1909) *A Manual of Practical X-ray Work*. Rebman.
Benson, Lord (1992) The Professions. Official Report 5th Series. Parliamentary Debates 1992–93, Lords Vol. 538 Gun 15 1992 – July 9 1992, pp. 1208–1210.

Blane, D. (1986) Health professions. In Patrick, D.L., Scambler, G. (eds) *Sociology as Applied to Medicine*, pp. 213–220. London: Baillière Tindall.
Blane, D. (1991) Health professions. In Scambler, G. (ed.) *Sociology as Applied to Medicine*. London: Baillière Tindall.
Burrows, E.H. (1986) *Pioneers and Early Years. A History of British Radiology*. Alderney, Channel Islands: Colophon.
College of Radiographers (1994) Code of Professional Conduct. London.
Hickson, D.J., Thomas, M.W. (1969) Professionalization in Britain: a preliminary measurement. *Sociology* **3**: 37–53.
Hugman, R. (1991) *Power in Caring Professions*. London: Macmillan Press.
Illich, I. (1976) *Limits to Medicine*. Harmondsworth: Penguin.
Larkin, G.V. (1978) Medical dominance and control: radiographers in the division of labour. *Sociological Review* **26**: 843–858.
Larkin, G.V. (1983) *Occupational Monopoly in Modern Medicine*, p. 64. London: Tavistock.
Larson, M.S. (1977) *The Rise of Professionalism: A Sociological Analysis*. Berkeley: University of California Press.
Millerson, G. (1964) *The Qualifying Associations: A Study in Professionalization*. Routledge and Paul.
Moodie, I. (1970) *50 Years of History*. London: The Society of Radiographers.
Paterson, A. (1995) *Role Development – Towards 2000. A Survey of Role Developments in Radiography*. London: College of Radiographers.
Radiographers Board (1995) *Statement of Conduct*. Council for Professions Supplementary to Medicine. London.
Scott-Angell, N., Chalmers, R.J. (1992) Results of a survey on ultrasound examinations. *Radiography Today* **61(662)**: 34–37.
Society of Radiographers (1990) *Amended Memorandum and New Articles of Association of the Society of Radiographers Limited*. London: Society of Radiographers.
Wilensky, H.L. (1964) *Industrial Society and Social Welfare: The Impact of Industrialisation on the Supply of Social Welfare Services in the United States*. New York: Russell Sage Foundation.

2
Breast Screening – The Flawed Paradigm?

Derek Adrian-Harris

INTRODUCTION

The National Health Service breast screening programme (NHS BSP) was introduced in the UK following the recommendations of the Forrest Report in 1986. The major belief was that early detection of breast cancer would improve prognosis such that by the end of the century there would be a 30% reduction in deaths from this disease. Government predictions are that 16 000 women in the UK will die from breast cancer each year (Cumberledge, 1994) and, according to the cancer charity Tenovus, women living in Britain are more likely to contract breast cancer than to win ten pounds in the National Lottery.

The attractions of being able to reduce the impact of this killer disease are obvious, and mammography screening has become a big industry. In the UK there are, according to the Department of Health, 104 screening services, employing 3% of the NHS diagnostic radiography workforce, and probably the equivalent of 15 full time physicists and 150 consultant radiologists. Figures from their own data returns suggest that in the year ending April 1994 more than 1.2 million asymptomatic women between the ages of 50 and 64 years responded to the three-yearly invitation to be screened in these centres. Using the accepted figure of £28 per examination this is a cost to the NHS of more than £33 m, although more usually quoted by government sources as £27 m. Any costings for the Breast Screening Programme (BSP) are somewhat hazy because the cost for repeats due to technical failure are not included, and follow-up procedures arising from positive findings are a charge to the acute sector clinics to which the women are referred. Breast screening is well publicized and seems to be welcomed by, and popular with, both women and health promoters. Frequently the subject of encouraging and evangelical articles in the press, especially women's magazines (see for example *Marie Claire*, September 1995), it is often offered as an optional extra for health checks provided by the private medicine sector. As recently as June 1995, the popular press was hailing a significant fall in deaths from breast cancer, basing this contention on their misunderstanding of a letter printed in *The Lancet* from the Cancer Epidemiology Unit at Oxford (Beral et al. 1995).

Nonetheless, and in spite of its popularity with, and undoubted support from, some members of the medical and allied health professions, there has been a constant suggestion that mammography in general, and the NHS BSP in particular, is not effective. Skrabanek (a member of Forrest's working group) wrote in 1988 that because the report had been published as a consensus document, the case against setting-up a breast screening programme had never been publicized as the minority dissenting members of the working party were bound to secrecy. Within a month of the Beral letter, a paper in *The Lancet* (Wright and Mueller, 1995) described how it had been calculated that the cost of one saved life attributable to breast screening was over half a million pounds. They stated that 'the allocation of limited resources . . . must be based on a critical analysis of benefits, harm and cost . . . the benefit is marginal, the harm caused is substantial and the costs incurred are enormous, we suggest that public funding for breast cancer screening in any age group is not justifiable.'

The purpose of this chapter is to explore the evidence pertaining to the effectiveness of the NHS BSP in saving women's lives. Was it, as some cynics have suggested, a sop to women voters in an effort to secure a further election victory for an unpopular government? Secondary questions arise, including:

- Were the techniques and assumptions used by Forrest valid?
- Has biochemistry become more effective than mammography as a screening tool?
- Is breast cancer a good candidate for an NHS screening programme?

FORREST REVISITED

Forrest was influenced by the outcomes of large screening programmes which had started in the early 1980s in two centres in Sweden and a third in the USA. A very large number of women had been involved (almost 0.25 million) and the data seemed statistically significant. However, there was inconsistency between the trials relating to the number of projections per breast and to the age range of the participants. A concurrent study was nearing completion in Edinburgh, but the Forrest committee did not wait for its findings to be published. With the advantage of hindsight it seems strange that the committee did not extend the period of deliberations to include the Edinburgh findings, especially as Forrest was then a professor of surgery in that city. Perhaps it was confident that Edinburgh would confirm the apparent trends from the USA and Sweden or perhaps there were external pressures to publish without delay. The data generated by these studies was deemed to be applicable to the UK population even though it is accepted that the frequency of occurrence of this disease differs amongst ethnic groupings. The article by Wright and Mueller emphasizes the point that of six major breast screening trials only two (New York and Malmo) showed benefits in reducing death rates. Indeed, the last of the series (Canada) recorded a slightly raised mortality from the disease amongst the women screened. V. Svoboda (1995, personal communication) a radiotherapist is convinced by observations drawn from his practice that climate is a major influence in triggering the disease, and his belief is to some extent confirmed by the UK maps of cancer incidence. Acceptance of such a theory would support the Canadian experience and might explain why a trial involving 30 000 women in Holland failed to demonstrate any statistically significant gain from breast screening. This could be a major variable that was not considered by Forrest. Quite possibly, a new group constituted to review mammography would use measures additional to those used in 1986 and justify their proposals in the light of both positive and negative evidence arising from all presently known trials.

In conventional medicine the large number random trial has been the research tool of choice and in the late 1980s this was certainly the case. Another measure has evolved which puts greater emphasis on clinical rather than statistical significance. Sackett et al. (1991) is one of the proponents of the 'number needed to treat' (NNT) concept. This refers to the number of treatments required to save one life or incur one significant side effect. For the purposes of this paper, however, the concept applies with equal validity to diagnostic investigations.

Sackett identifies the NNT measure as having three advantages beyond its clinical usefulness. Firstly, it provides a measure of the effort required by both physician and patient to achieve some treatment target. Secondly, it provides a measure of cost for both the treatment and its monitoring, and finally it can predict the frequency (or time scale) with which organ damage or death will occur. Wright and Mueller's findings could be considered in the light of an NNT concept, where from the six screening programmes, between 7086 (New York) and 63 264 (Malmo) women would need to be examined before they can claim one life saved. In Canada the number needed is infinite. The argument for introducing the Forrest Report was that there would be a 30% reduction in deaths from breast cancer, but this does not seem to reconcile with the Canadian findings above. In fact it is consistent but the explanation seriously weakens the attractions of any breast screening programme. The reduction in mortality is *relative* rather than *absolute*. The two Scandinavian studies produced relative reductions of 21–30%, but the absolute mortality reduction amongst the women screened was 0.05–0.14%.

The argument is as follows: consider two populations each of 10 000 women, where the first group is to be screened and the second group is to act as a control. Fifteen deaths could be predicted amongst the control group. Within the screened population the detection and treatment of breast cancers would reduce the death rate in this group to 11 persons which is a reduction of 30% (as predicted by Forrest). However, the absolute saving is only 4 lives from the 10 000 subjects who formed the screened population.

Across the six programmes surveyed by Wright and Mueller's colleagues, about 5% of examinations produced a positive finding. After follow-up investigations and, presumably, much distress to the women and families concerned, between 80% and 93% of these alarms were found to be false positives. Conversely, up to 15% of the examinations were false negatives which became clinically apparent within one year of the wrongly reassuring screening. In other words, the justification of mammography screening because of the reassurance it provides (when 'nothing is found') is fallacious.

Whilst Wright and Mueller conclude that 'the public funding of breast screening is unjustifiable', an American contemporary, Gofman (1995), has claimed that during the period 1920 to 1960 medical radiography is responsible for up to 75% of the occurrences of breast cancer in the USA. Radiobiologists have always acknowledged the risk–benefit philosophy inherent in radiological practice. What is new is the suggestion that X-rays trigger breast cancer.

Acceptance of Gofman's claims, allied to Wright and Mueller's evaluations and the NNT approach, would bring very close the immediate cessation of asymptomatic mammography. The National Radiological Protection Board (NRPB), amongst others, has been quick to point out that not only are Gofman's claims based on his own dose calculations, but they are dramatically different from the perceived norms of dose and risk. However, it must be recognized that the claims originate from a respected award winning scientist and Berkeley professor. As radiographers become aware of the issues from the fluoroscopy of tuberculosis patients (Mackenzie, 1965) they may be less ready to dismiss such unpalatable findings without a second and a third thought.

BIOCHEMISTRY, A NEW WAY FORWARD?

Can biochemistry offer a more elegant vehicle for breast cancer screening? If an appropriate blood chemistry test was available it might be cheaper; more accurate (the incidences of false positives and false negatives occurring within mammography have already been discussed), and arguably more acceptable to the target screening population, some of whom complain of embarrassment and discomfort. Clearly, a blood test carries no risks from radiation, and in the past two years evidence has become available which suggests that blood chemistry may be a fruitful avenue for exploration in the attempt to reduce the mortality of breast cancer.

A team of researchers from Utah, led by Miki, has identified genetic predictors of the disease which are apparent in DNA. These researchers believe that a population screening programme could be based on DNA studies to identify the presence of the predictor mutation (Miki et al., 1994). Similar findings have been described by a group from Michigan (Weber et al., 1994). A contribution from Seattle (Malins et al., 1995) asserts that cancerous breast tissue can be identified with a sensitivity and specificity of 83% by the application of Fourier transformation infra-red spectroscopy. They conclude 'progressive structural changes in the DNA of the normal female breast, leading to a pre malignant cancer-like phenotype ... are the basis for predicting its occurrence at early stages of oncogenesis'. From France, it has been reported that the measure of oestrogen receptors in RNA is a good prognostic indicator in women with breast cancer (Gotteland et al., 1994). There seems to be sufficient evidence to anticipate that laboratory tests provide a viable alternative screening mechanism to mammography. At least, further studies of the potential role of biochemical tests in the detection of breast cancer should be undertaken.

BREAST CANCER, A DISEASE SUITABLE FOR SCREENING?

Even if the doubts about mammography were to be resolved, the basic question would still remain. Is cancer of the breast a worthy candidate for a screening programme? Editorials in the *British Medical Journal* (1988) and *The Lancet* (1993), suggest that some clinicians have doubts. In September 1995, Professor Michael Baum resigned from the NHS breast cancer advisory group because 'the X-ray tests done on more than 1 million women a year are not worth doing'. Although not the first, this is the biggest attack yet on the NHS BSP. Skrabanek wrote in 1985 'The evidence that breast cancer is incurable is overwhelming ... the philosophy of breast cancer screening is based on wishful thinking that early cancer, is curable cancer.' In support of this assertion he quoted over 70 references but, in summary, he posited that there is little evidence to suggest that breast cancer mortality has improved over the past 150 years, and that large tumours (> 6 cm) seem to have a better prognosis than small ones. Most tumours detected by mammography have, at the time of detection, already undergone about 29 doublings in size. By 40 doublings, the tumour load borne by the patient is likely to prove fatal. As many as 77% of tumours grow from sub-threshold size (by mammography) to being clinically significant in less than 12 months, and 90% of tumours have metastasized by the time they have reached a diameter of 6 mm. By inference, such data must have implications for the justification of a three year interval between mammograms. However, Skrabanek whose criticisms are summarized above, is not a lone voice. Berg (1995), a pathologist, described how there are more than 20 different breast cancer histologies, and within these variants lies the true determinant of both curability and prognosis.

The most recent figures from the UK (OPCS, 1992, Mortality statistics) show that 13 663 people including 92 men, died of breast cancer. It should not go unremarked that the 1992 figures are 20% fewer than the Cumberledge predictions of 16 000 deaths per annum. More up-to-date figures will soon be available and have been hinted at by Beral. Care is needed on two counts. Firstly, the counting base has been changed since 1992 and, secondly, errors may be introduced because classifications are made from information on death certificates. For example, a patient who is terminally ill with cancer dies in a road traffic accident (RTA) on the way to a hospice. The probable recorded reason for death would be trauma, even though the victim might have died the next day from cancer had the RTA not intervened. Alternatively the individual's weakened state due to cancer might lead to death from injuries which sturdier people would probably have survived. Even with this caveat, the OPCS data show that 63% of breast cancer deaths occur in those older than 65 and that the most common age group to die from this cause is that aged 70–74 (almost matched by the 75–79 group). A histogram of breast cancer mortality by age is shown in Figure 2.1.

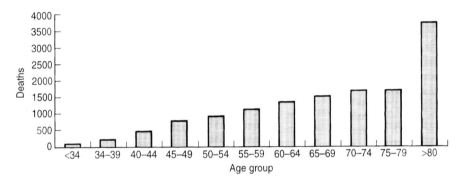

Figure 2.1 Mortality by age–breast cancer. Based on data from OPCS (1994).

But what, from the government's health perspective, is so special about women? Why not a prostate screening service? During the same year, almost 9000 men died of prostatic cancer, most commonly at the same age as the women (the Tenovus institute also equated men's chances of succumbing to prostate cancer to be greater than their chances in the National Lottery – see first paragraph). This may be irrelevant, since the biggest causes of UK deaths are heart and lung disease. A screening programme for heart disease, aneurysm detection or lung disease would be gender free and have more impact on UK mortality rates.

A histogram of some of the OPCS data relating to causes of death is shown in Figure 2.2.

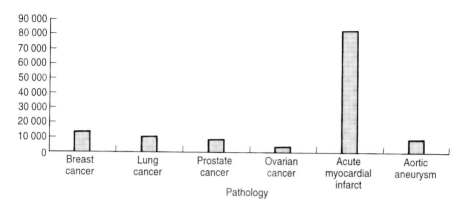

Figure 2.2 Common causes of death. From OPCS (1994).

CONCLUSIONS

It would seem that there is evidence to believe that the present NHS BSP is not saving women's lives. It may even be endangering them and certainly it is causing distress. Future developments in biochemical testing may generate a better screening tool but, even so, doubts remain about the belief that 'early cancer, is curable cancer'. The breast screening programme is offered to only half the population – women. Potentially, at least, there are diseases affecting both genders which are better candidates for screening programmes. The desirability of saving life and avoiding disfiguring illness is not in doubt but the efficacy of the present breast screening programme remains unclear. Surely, it must be time to question its continued support and funding in the apparent absence of value for money or other tangible outcome measures.

As a profession, radiographers have a responsibility to examine their practice periodically. During her year as president of the Society and College of Radiographers, Boutle (1995) said, 'Critical examination of traditional or accepted practice could have a profound effect on the well-being of untold numbers of people and should not be ignored'. Medicine and technology have moved on and radiography still flourishes. The population's interest may be better served by allowing colleagues to develop and substitute new techniques and the opportunity to engage in discussion should be welcomed. Modern medicine in general and the profession of radiography within it are too big to need sacred cows. As a group, radiographers should seek to determine the efficacy of mammography screening whilst at the same time expecting medical researchers to provide the evidence in support of their belief that early diagnosis saves lives.

COMMENT   Richard Price, *Hatfield*

The Forrest Report (1986) concluded that 'the only way substantially to reduce the number of deaths from the disease [cancer of the breast] is to detect it before the patient presents with symptoms'. On the available evidence mammography was the only proven method of detecting breast cancer and reducing mortality in women over 50 years of age. In response to the report the government, in 1987, established the NHS Breast Screening Programme (NHS BSP) which consists of mammography screening, assessment, biopsy and histological examination followed by treatment. The effectiveness of this screening programme is now being challenged.

Derek Adrian-Harris in this chapter has reviewed some of the evidence now available and concluded; ... there is evidence to believe that the present NHS BSP is not saving women's lives.' Some evidence is irrefutable; in 1992, 13 663 women in England and Wales died as a result of breast cancer (Select Committee on Health, 1995). This was 4.8% of female deaths and more than were caused by any other cancer or disease excepting coronary heart disease and cerebrovascular disease. Unfortunately, the incidence of the disease in the UK is increasing and the Breast Screening Programme has certainly contributed to the increase in the recorded incidence by detecting more cancers at an earlier stage than if screening had not been introduced. But surely this would be expected of a screening service established for the very purpose of early detection. Whilst Adrian-Harris questions continuing support for breast screening it is hardly the time to take decisions when there is insufficient evidence available. As was reported to the House of Commons Health Select Committee in 1995, there is no expectation to see any reduction in mortality as a result of screening until 1997. Wald et al. (1994), stated that screening programmes have not been underway for sufficient time to assess the effect on breast cancer mortality. Whilst the true situation concerning the NHS BSP has yet to emerge, there is additional evidence cited by Wald et al. (1994) which is that:

> ... screening for breast cancer by inviting women for periodic mammographic examination, followed by diagnosis and treatment as necessary, leads to a significant reduction in breast cancer mortality.

The quotation sums up the intention as set out in the Forrest Report (1986) which had found that developments in treatment had not produced significant increases in survival. Hence, the recommendation to make every effort to detect the disease before the onset of symptoms. It will be the outcome of treatment of the screen-detected cancers that will be used as the indicator of the effectiveness of the screening programme.

Clinical management is therefore critical, but it is the screening programme which has provided the focus for treatment of breast cancer. Quality assurance guidelines for surgeons were produced and indeed were the first in the world for surgeons working in a national breast screening programme.

A quote from Professor Roger Blamey (1992) chairman of the British Association of Surgical Oncology National Surgical Co-ordinators Group stated

> We can now say to women in the programme, not only can we cure your cancer by early detection, we can do so with the minimal surgery necessary so that you are left with a good cosmetic result.

But it is perhaps the treatment that remains the main cause for concern. The Select Committee on Health published its third report in July 1995. This reviewed breast cancer services, highlighted within the report were the findings of the Yorkshire Cancer Organisations. The findings described differences in treatment over a small geographical area. It was reported that a woman in one town was 30% more likely to have her whole breast removed than a woman living in a nearby town. It was considered that the difference was unlikely to be the result of differences in the stage of the disease at presentation. There was also cause for concern in the type of adjuvant therapy being offered in Yorkshire. The Health Committee expected to find minor variations in treatment but found it hard to accept the differences reported in Yorkshire and called for the publication of data on a national basis. As the Health Select Committee noted in commenting upon the comparatively higher incidence and lower mortality of the disease in the USA where screening has been available for longer

> The most worrying explanation, however, would be that the UK's poor survival rates may be due, at least in part, to poor treatment of the disease.
>
> Health Select Committee (1995, p. x)

There are claims that patients referred to centres of excellence receive a higher quality of care and expertise than those referred to a nonspecialized unit the future of breast cancer care is through specialist breast units and we feel that this is the way to reduce breast cancer mortality in the UK.

Health Select Committee (1995, p. xii)

Surely, the establishment of specialist centres nationally is the priority of a national service?

Other concerns in the NHS BSP that have been acknowledged, such as false positive recalls, the associated anxiety and interval cancers are being addressed by recent changes to the programme. But at present, mammographic screening is the only method of detecting small nonpalpable cancers.

It must be recognized that the publicity generated by the programme has greatly increased public awareness and compliance, which in itself will lead to earlier detection and treatment. The role of the media, especially women's magazines, are recognized for their increasing contribution to awareness of the disease. As for the politicians, an objective stated in *The Health of the Nation* (1991) is to reduce breast cancer. The target for breast cancer was stated thus.

> To reduce the death rate for breast cancer in the population invited for screening by at least 25% by the year 2000 (from 95.1% per 1 000 000 in 1990 to no more than 71.3% per 100 000).

Who will take the responsibility for terminating the current programme? Surely, this is not the time to question its effectiveness when sufficient data will not be available until 1997 at the earliest.

The national screening programme has brought a focus and momentum upon which to capitalize. To withdraw the NHS BSP is not an option at the present time and the establishment of more specialist treatment centres with multidisciplinary teams of experts must be a priority. The attack on breast cancer must continue on all fronts; the screening programme is only a part of that attack; research remains fundamental to understanding, detection and treatment of this horrific disease.

References

Beral, V., Hermon, C., Reeves, G., Peto, P. (1995) Sudden fall in breast cancer deaths. [letter]. *Lancet* **345**: 2406.

Berg, J.W., Hutter, R.V. (1995) Breast cancer. *Cancer* **75**: 257–269.

Blamey, R. (1992) Quotation from *Rad Magazine*, August 30.

Boutle, R. (1995) Foreword for *Radiography* **1**: 3.

Cumberledge, Baroness (1994) NHS Breast Screening Programme Review. NHS.

Editorial (1988) Controversy over mammography. It should save lives. *British Medical Journal* **297:** 932–933.

Editorial (1993) Breast cancer: have we lost our way? *Lancet* **341:** 343–344.

Forrest Report (1986) *Breast Cancer Screening.* Report of a Working Group. London: HMSO.

Gofman, J. (1995) *Preventing Breast Cancer; the Story of Major, Proven Preventable Cause of this Disease.* San Francisco: CNR Books.

Gotteland, M., May, E., May-Levin, F. et al. (1994) Estrogen receptors (ER) in human breast cancer. *Cancer* **74:** 864–871.

Mackenzie, I. (1965) Breast cancer following multiple fluoroscopies *British Journal of Cancer* **19:** 1–9.

Malins, D.C., Polissar, N.L. et al. (1995) The etiology and prediction of breast cancer. *Cancer* **75:** 503–517.

Miki, Y., Swenson, J., Shattuck-Eidens, D. et al. (1994) A strong candidate for the breast and ovarian cancer susceptibility gene BRCA 1. *Science* **266:** 7 Oct.

Office of Population Censuses and Surveys (OPCS) (1994) *UK Mortality Statistics 1992.* London: HMSO.

Sackett, D.L., Haynes, R.B., Guyatt, G.H., Tugwell, P. (1991) *Clinical Epidemiology,* 2nd Edition. Boston, Mass.: Little Brown.

Select Committee on Health (1995) *Breast Cancer Services* (Third Report). London: HMSO.

Skrabanek, P. (1985) False premises and false promises of breast cancer screening. *Lancet* August 10, 316–319.

Skrabanek, P. (1988) The case against breast screening. *British Medical Journal* **297:** 971–972.

The Health of the Nation; A Strategy for Health in England (1991). London: HMSO p. 67.

Wald, J.J., Chamberlain, J., Hackshaw, A. (1994) European Society of Mastology consensus: report of the evaluation committee. *British Journal of Radiology* **67:** 925–933.

Weber, B.L., Abel, K.J., Brody, L.C. et al. (1994) Familial breast cancer. *Cancer* **74:** 1013–1020.

Wright, C.J., Mueller, C.B. (1995) Screening mammography and public health policy: the need for perspective. *Lancet* **346:** 29–32.

3

Computed Tomography – A Spiralling Challenge in Radiological Protection

Paul C. Shrimpton

INTRODUCTION

The widespread use of X-rays in medicine ensures that such examinations represent by far the largest man-made source of population exposure to ionizing radiation. In the UK, as in most other developed countries, diagnostic radiology provides about 90% of the radiation dose to the population from all artificial sources and about 13% of that from all sources including natural background radiation (Hughes and O'Riordan, 1993). Notwithstanding the significant benefits to patients from such medical exposures, the principal concern in radiological protection is the reduction of *unnecessary* exposures by means of adequate clinical justification and optimization of patient protection (National Radiological Protection Board, 1993). These guiding principles are also enshrined in legislation, which relate both to the equipment (Ionising Radiations Regulations, 1985) and the procedures used (Ionising Radiation (Protection of Persons Undergoing Medical Examination or Treatment) Regulations, 1988).

Over the last two decades in particular, there has been a growing awareness through surveys of practice of the levels of dose received by patients (Wall et al., 1980; Shrimpton et al., 1986) and of the potential and need for dose reduction (NRPB, 1990). During this time, there has also been a steady proliferation of the revolutionary x-ray imaging technique of computed tomography (CT) following its introduction into clinical practice in 1972. CT provides high quality cross-sectional images of the patient and increasing application of this modality in the diagnosis and assessment of cancer and other pathological conditions has undoubtedly made a substantial impact on patient care. However, the concomitant impact on population exposure was not fully appreciated until a national survey in the UK in 1989 established that the typical levels of patient dose from CT are relatively large compared with those for many conventional x-ray examinations of similar regions of the body, as illustrated in Table 3.1 (Shrimpton et al., 1991a,b; Jones and Shrimpton, 1991). Accordingly, CT procedures were estimated to represent about 2% of the total of all x-ray examinations, yet account disproportionately for about 20% of the resultant collective dose of some 20 000 man Sv in 1989.

Table 3.1 Levels of patient dose from x-ray examinations in the UK

Examination	Typical effective dose (mSv)	
	CT	Conventional
Head	1.8	0.1
Thoracic spine	5.8	1.1
Chest	8.3	0.04
Abdomen	7.2	1.4
Lumbar spine	3.6	2.2
Pelvis	7.3	1.0
Intravenous urography	–	4.6
Barium meal	–	4.6
Barium enema	–	8.7

This assessment firmly established CT as a major source of exposure from diagnostic X-rays and a technique worthy of particular attention in terms of radiation protection for patients. The present challenge concerns how much of this exposure may be *unnecessary*.

CONTINUING GROWTH IN CT PRACTICE

CT heralded the beginning of the exciting era of digital radiology and has, perhaps, sometimes been perceived inherently to involve low patient doses in view of the use of state-of-the-art technology and powerful computers. Indeed, a survey in one English county of specialist clinicians (other than radiologists) who worked with X-rays revealed that 20% did not realize that CT scanning involved ionizing radiation (Keal, 1991). Expansion in CT practice, initially at least, has probably occurred in ignorance of the relatively high doses involved. This situation has been fuelled in part by the difficulties in adequately characterizing patient exposure arising from the complex pattern of irradiation during CT scanning, which is quite different from that during conventional examinations. Comprehensive assessment of the radiological risk from CT requires knowledge of organ doses and this has necessitated the development of specific dosimetry techniques; these involve the measurement free-in-air of the dose on the axis of rotation of the scanner and normalized organ dose data determined for slabs of a mathematical anthropomorphic phantom using Monte Carlo techniques (Shrimpton et al., 1991b; Jones and Shrimpton, 1991).

Data from surveys of CT practice in other developed countries have indicated similar patterns to that in the UK (Shrimpton and Wall, 1995). The most recent assessment of medical exposures by the United Nations Scientific Committee on the Effects of Atomic Radiation (UNSCEAR, 1993) suggests a contribution from CT of about 14% of the total collective dose from x-ray examinations on a global scale, although this varies markedly between countries with differing levels of health care, as categorized on the basis of number of physicians per head of population; the average contributions are 18% for countries in the top level and only 2% for those in the three lower levels.

On the basis of UK practice for 1989, the collective dose from each CT scanner equates typically to about 20 man Sv per year. This is, for example, similar to the annual collective dose to the population arising from routine discharges to the environment from the whole of the UK nuclear industry (Hughes and O'Riordan, 1993). Further analysis of CT practice at two particular hospitals in 1991 has indicated remarkably different trends in collective dose relative to data for 1989 (Crawley and Rogers, 1994). At the first, where magnetic resonance imaging (MRI) was available and where dose reduction measures had been implemented, there was a 30% reduction in patient numbers and a 50% reduction in collective dose to a level of 4 man Sv. At the second hospital, however, where MRI was not available and where the introduction of dose saving strategies had been delayed, there was a 25% increase in patient numbers and a 100% increase in collective dose to a level of 48 man Sv.

The number of CT scanners in use in the UK has continued to

increase steadily since the survey in 1989, as shown in Figure 3.1. On this evidence, the contribution from CT may have risen to nearly a third of the total collective dose from medical X-rays (Shrimpton and Wall, 1995), with an increase in absolute terms of over 3300 man Sv between 1989 and 1995.

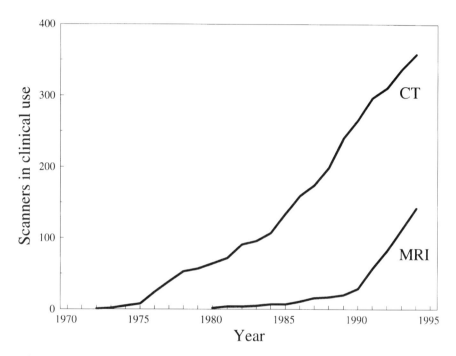

Figure 3.1 Growth in provision for CT and MRI in the UK. Data from the Medical Devices Agency (DEP) (personal communication, 1995) and the Royal College of Radiologists (1992).

Figure 3.1 also indicates the steady growth in the number of MRI scanners in the UK, broadly parallel but offset in time by about a decade relative to data for CT. MRI is now applicable in a wide range of clinical circumstances and, since it does not involve ionizing radiation, the Royal College of Radiologists (RCR, 1995) has recommended that MRI should be preferred to CT where both imaging modalities would provide similar information and when both are available. Sales in MRI equipment in the UK were about twice those for CT in 1993 (Ward, 1994), although the total provision for MRI is still some way below the target level of about 225 units that the RCR (1992) suggests is necessary for the effective and efficient utilization of the technique. This level would in principle also allow significant reduction in population exposure to medical X-rays.

It is likely, therefore, that in the short term at least practice in CT will continue to increase. One reason, perhaps, for renewed and continuing interest in CT is the introduction of the technique of helical (spiral) scanning in 1989. This provides significant clinical advantages by allowing the rapid acquisition of image data over large volumes of the patient. The innovation has considerably extended the usefulness of CT so as to rechallenge MRI, which remains a relatively slow technique. Image quality and patient dose in helical CT are broadly similar to those for conventional slice-by-slice imaging when equal or equivalent scan parameters are chosen (Kalender, 1993). However, the speed and convenience of helical scanning surely offers a potential for increased doses by encouraging the imaging of larger volumes of the patient.

GENERAL TRENDS IN DOSE REDUCTION

Notwithstanding the pattern for CT, conventional x-ray imaging has been a focus for dose reduction since a national survey demonstrated significant variations in patient dose for a given type of examination and raised concerns over unnecessary x-ray exposures (Shrimpton et al., 1986). A subsequent joint report by NRPB and the RCR reviewed a wide range of methods for reducing doses to patients and made recommendations relating to justification, optimization of the procedures and equipment, and staff training (NRPB, 1990). Overall, it was estimated that a reduction of nearly one-half was possible in the collective dose from conventional x-ray examinations, without adversely affecting patient care. In particular, it was suggested that at least 20% of all x-ray examinations were clinically unhelpful in terms of providing information useful to patient management.

As a practical aid to the optimization of protection, NRPB have specified national reference doses for some common types of examination, with the recommendation that hospitals should aim to achieve mean doses below these levels (NRPB, 1993). Present levels are based on third quartile values for the distributions of mean doses at individual hospitals participating in the national survey in the 1980s. They are expressed in terms of the directly measurable dose quantities: entrance surface dose per radiograph and dose–area product per examination.

Such reference levels have the function of local investigation levels for x-ray departments and should be applied with flexibility to allow higher doses where indicated by sound clinical judgement. Otherwise changes should be made to technique or equipment to reduce doses below the reference level without compromising image quality.

In order that such a system of national and local reference doses can be successful in promoting dose reduction, it is necessary to include regular patient dose monitoring as a part of routine quality assurance. This has been facilitated by publication of a protocol for patient dosimetry in diagnostic radiology, setting out methods nationally agreed between NRPB, the College of Radiographers (COR) and the Institute of Physical Sciences in Medicine (IPSM) (IPSM/NRPB/COR, 1992). X-ray departments are encouraged to participate in a system for the national collation by NRPB of local patient dose data. This will allow central assessment of the impact of patient protection measures on diagnostic radiology in the UK, and for the national reference doses to be extended and periodically revised so that they continue to fulfil their purpose of promoting optimization of patient protection.

A preliminary analysis of the national database has indicated an encouraging trend towards lower patient doses for simple radiographic examinations, with an estimated saving in annual collective dose of about 1000 man Sv (Wall, 1994). The database is continuing to grow steadily in both the amount and range of data, and by mid 1995 included approximately 30 000 measurements from over 1000 x-ray rooms at over 300 hospitals.

Such an integrated approach to dose reduction has clearly proved effective for conventional examinations and suggests that a similar system would be worthwhile for CT.

QUALITY CRITERIA FOR CT

The national survey of CT practice has already indicated potential scope for improvement in the optimization of protection for patients undergoing such examinations. Significant variations were observed between individual scanners in the typical dose for a given type of procedure; factors of 10–40 over all scanners and factors of 5–20 when each scanner model was considered separately (Shrimpton et al., 1991b). Moreover, only 7% of users reported the routine performance of any type of dose assessment (Shrimpton et al., 1991a). In order to promote better control of patient dose from CT, the Board has published for the various professional groups involved with CT general recommendations concerning protection of the patient (NRPB, 1992). This advice relates to clinical practice, the design and maintenance of CT equipment, and staff training. It includes recommendations for the development of a dosimetry protocol for CT and examination-specific reference levels of dose.

In response to growing concern, a study group was established in 1994 by the Commission of the European Communities to develop quality criteria for CT, complementary to similar initiatives for conventional x-ray examinations and paediatric radiology. The criteria will include recommendations on good practice for general types of CT procedure and guidelines on patient dosimetry. The advice concerns justification, good imaging technique, image viewing, and image and dose criteria for the evaluation of performance. The development of appropriate reference doses is proving particularly complex since this requires quantities that provide a meaningful indication of patient exposure for all types of scanner and technique, yet are simple to measure or determine. Draft criteria are scheduled for publication in 1996.

In the meantime to encourage more widespread dosimetry, the methods of the national survey and the results of Monte Carlo calculations of normalized organ doses for CT have been made available in a software report (Jones and Shrimpton, 1993).

CONCLUSION

CT represents an increasingly important source of exposure to medical X-rays. There is an urgent and continuing need for close scrutiny of practice so as to eliminate all unnecessary exposures from an otherwise valuable clinical tool. Any reductions in patient dose from conventional x-ray examinations that have been achieved by coherent initiatives over the last few years will probably have been offset by significant increases in collective dose arising from continuing growth in CT.

Pending the formal development of similar dose reduction strategies for CT, particular attention should be paid to the justification of each examination (RCR, 1995) and to the optimization of patient protection. Fundamentally, patient dose for an examination is proportional to the radiographic exposure (mA s) per slice, the number of slices and the slice width, and control of these parameters provides the practical means for dose reduction. Helical scanning in particular offers the potential for increasing the complexity of examinations and also the patient dose. There is a joint challenge and responsibility on radiologists, radiographers and physicists to ensure the effective and efficient use of CT with an associated containment of unnecessary exposures.

References

Crawley, M.T., Rogers, A.T. (1994) A comparison of computed tomography practice in 1989 and 1991. *British Journal of Radiology* **67**: 872–876.

Hughes, J.S., O'Riordan, M.C. (1993) Radiation exposure of the UK population – 1993 review. Chilton, NRPB-R263. London: HMSO.

Ionising Radiations Regulations (1985) SI 1333. London: HMSO.

Ionising Radiation (Protection of Persons Undergoing Medical Examination or Treatment) Regulations (1988) SI 778. London: HMSO.

IPSM/NRPB/COR (1992) *National Protocol for Patient Dose Measurements in Diagnostic Radiology*. Chilton: NRPB.

Jones, D.G., Shrimpton, P.C. (1991) *Survey of CT Practice in the UK* Part 3: Normalised organ doses calculated using Monte Carlo techniques. Chilton, NRPB-R250.London: HMSO.

Jones, D.G., Shrimpton, P.C. (1993) Normalised organ doses for x-ray computed tomography calculated using Monte Carlo techniques. Chilton, NRPB-SR250.

Kalender, W.A. (1993) The physical principles and technical performance of spiral computed tomography. *Medical Physics* **20(3)**: 940–941.

Keal, R. (1991) Use of ionising radiation by non-radiologists. *RAD Magazine* **17(197)**: 13–14.

NRPB (1990) Patient dose reduction in diagnostic radiology. *Doc. NRPB* **1**, No. 3.

NRPB (1992) Protection of the patient in x-ray computed tomography. *Doc. NRPB* **3**, No. 4.

NRPB (1993) Medical exposure: Guidance on the 1990 recommendations of ICRP. *Doc. NRPB* **4**, No. 2, 43–74.

RCR (1992) *Provision of Magnetic Resonance Imaging Services in the UK*. London: RCR.

RCR (1995) *Making the Best Use of a Department of Clinical Radiology*, third edition. London: RCR.

Shrimpton, P.C., Wall, B.F. (1995) The increasing importance of x-ray computed tomography as a source of medical exposure. *Radiation Protection Dosimetry* **57(1–4)**: 413–415.

Shrimpton, P.C., Wall, B.F., Jones, D.G. et al. (1986) *A National Survey of Doses to Patients Undergoing a Selection of Routine x-ray Examinations in English Hospitals*. Chilton, NRPB-R200. London: HMSO.

Shrimpton, P.C., Hart, D., Hillier, M.C. et al. (1991a) *Survey of CT Practice in the UK*. Part 1: Aspects of examination frequency and quality assurance. Chilton, NRPB-R248. London: HMSO.

Shrimpton, P.C., Jones, D.G., Hillier, R.C. et al. (1991b) *Survey of CT Practice in the UK*. Part 2: Dosimetric aspects. Chilton, NRPB-R249. London: HMSO.

UNSCEAR (1993) *Sources and effects of ionizing radiation. Report to the General Assembly, with Scientific Annexes*. New York: UN.

Wall, B.F. (1994) National trends in patient dose. In Goldfinch, E.P. (ed.) *Proceedings IRPA Regional Congress on Radiological Protection, Portsmouth, June 1994*, pp. 121–124. Ashford: NTP.

Wall, B.F., Fisher, E.S., Shrimpton, P.C. et al. (1980) Current levels of gonadal irradiation from a selection of routine diagnostic x-ray examinations in Great Britain. Chilton, NRPB-R105. London: HMSO.

Ward, P. (1994) Health reforms create market uncertainties. *Diagnostic Imaging International* **10(8):** 13–20.

4
Technology and Changing Practice in Radiotherapy

Kate Westbrook

INTRODUCTION

Planning and delivery of radiotherapy treatment utilizes a wide range of technologies, not only in the radiation field but in computing and engineering. Changes in technology provide an opportunity for radiographers to address issues relevant to their clinical practice. Some of the key related issues are presented in Table 4.1 and it is these which are examined in this chapter.

Table 4.1 Issues relevant to the clinical practice of radiographers in the future

Radiographers' role
Radiotherapy departments of the future
Will technology change treatment techniques?
Will technology improve efficiency?
What new skills will radiographers require?
Training issues
Are there new technologies on the horizon?

RADIOGRAPHERS' ROLE

To be effective, radiographers require patient care skills together with technical expertise. In many centres, roles are already expanding rapidly as described by Paterson (1994). This chapter concentrates on technical issues but these must be considered alongside the developing patient care roles such as counselling.

Clearly, there is enormous technological growth and a number of recent innovations, such as multileaf collimators and electronic portal imaging, are available in a limited number of centres. It seems likely that such innovations will become the norm in radiotherapy departments within the coming decade and there is little doubt that the introduction of new technology will have a significant impact on work practices. Radiographers will be required to perform a whole range of complex tasks and develop new skills which will keep the profession at the forefront of cancer care and treatment. Indeed, the profession has the opportunity to respond and embrace the new technologies positively and enthusiastically.

The radiographer's role is likely to be extended as a result of Calman's report (1993) which recommended that junior medical staff are allowed adequate time for study. The implementation of the report along with the recommendations of the recent Expert Advisory Group on Cancer (1995) will provide the opportunity for radiographers to take an increased responsibility for technical aspects of radiotherapy treatment. In particular, treatment localization and verification are areas which should be considered.

RADIOTHERAPY DEPARTMENTS OF THE FUTURE

Before radiographers can address the impact of new technologies on working practice, it is worth considering which technologies will be most useful, both in treatment planning and on the treatment floor.

Table 4.2 shows the likely treatment planning options since localization in the treatment position requires either conventional simulation, computerized tomography (CT)/magnetic resonance imaging (MRI)/or ultrasound. Occasionally, more than one modality may be employed for an individual patient.

Table 4.2 Treatment planning equipment options

Simulator	3D computer planning
CT simulator	
CT scanner	
Ultrasound	
MRI scanner	

A number of manufacturers produce a CT simulator and others are currently developing an add-on CT scanning option for their simulators, both of which will allow CT data to be acquired in a wider range of treatment positions than can be achieved by a conventional CT scanner.

Open architecture MRI scanners are becoming available which will facilitate the use of imaging in the treatment position. However, it will require development in terms of tissue characterization and image linearity before it becomes an effective tool for planning.

Whether positron emission tomography (PET) should be included in the technologies required for treatment planning is more controversial and local decisions will have to be taken.

Three-dimensional (3D) computer planning systems will allow more complex treatment plans for conformal radiotherapy to be produced.

Figure 4.1 shows a simulator verification film of a right posterior

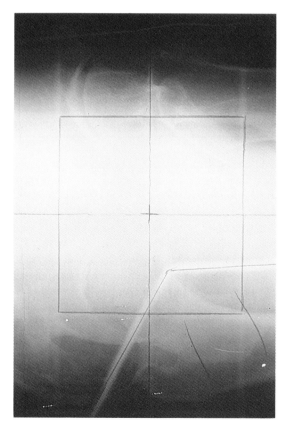

Figure 4.1 Simulator verification image.

oblique beam with shielding which forms part of a radical treatment for a patient with cancer of the prostate. Plate 1 shows a digitally reconstructed radiograph (DRR) of the same beam, which is produced by the planning system from the original planning CT data. Plate 2 shows the same DRR of the beam, but with the target volume superimposed. The availability of such images will give radiographers a much improved awareness of treatment volumes and surrounding anatomy than current two-dimensional systems, and provide a reference image for use in treatment verification.

Table 4.3 shows the options for equipment on the treatment floor. The bulk of the workload in any department is carried out on linear accelerators and therefore improvements in technology and applications will have the largest impact.

Table 4.3 Treatment floor equipment options

Linear accelerator	Photon and electron beams
	Multiple energies
Multileaf collimators	
Stereotactic collimation	
Megavoltage imaging	
Network	
Brachytherapy	

However, in the UK, brachytherapy occurs in most centres and there is an increasing trend towards high dose rate systems and the development of pulsed dose rate (Activity, 1994). These systems have a much wider clinical application and are sited within the treatment floor area rather than on oncology wards. Therefore the role of radiographers could develop in this field.

Table 4.3 does not include other particle therapies because it is likely that demand will be limited to specialist centres, although where they do exist radiographers will play an integral role in the development of techniques.

The following illustrations represent some of the more recent technological innovations.

Figure 4.2 is a view into the head of the Philips Multileaf Collimator (MLC) showing the 40 movable pairs of leaves which will allow complex treatment field shaping without the use of lead blocks.

Figure 4.3 shows the Philips Stereotactic Collimation system attached to one of their linear accelerators, which produces a fine pencil beam of radiation allowing very accurate treatment of brain lesions.

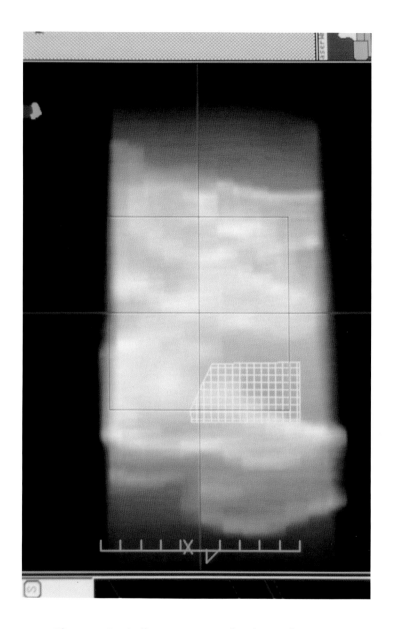

Plate 1 Digitally reconstructed radiograph (DRR).

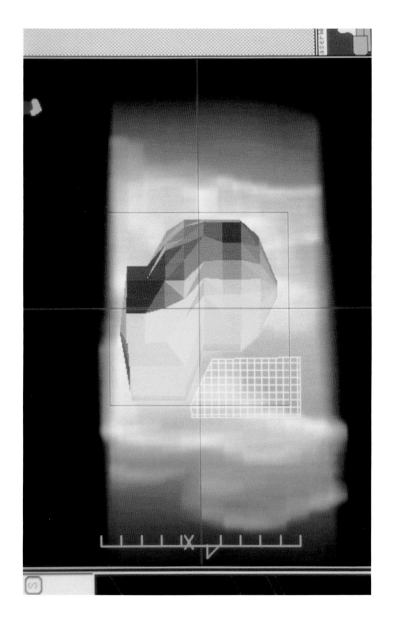

Plate 2 DRR with treatment volume superimposed.

Figure 4.2 Philips Multileaf Collimator (MLC).

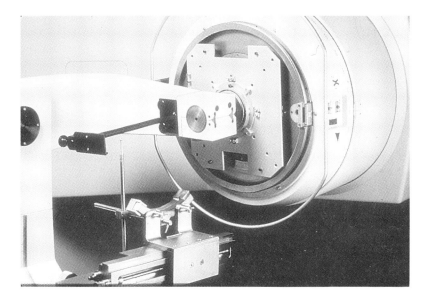

Figure 4.3 Philips Stereotactic Collimation System.

Inherent in such treatments is the requirement for a reproducible treatment set up which is achieved by utilizing a relocatable stereotactic frame. Although the use of stereotactic radiotherapy is likely to continue in a small number of specialized centres, radiographers should become involved in translating the immobilization techniques to other anatomical sites.

A typical electronic portal imaging (EPI) system which allows verification of a treatment field set up without the use of film is shown in Figure 4.4. A hard copy image can be produced if required using a video printer.

A schematic representation of the Philips Radiotherapy Network (RTNet) is shown in Figure 4.5 and illustrates how information can be transferred between nodes in the department, including the simulator. Equally, information can be relayed to and from a treatment planning system via the external computer interface, or data may be downloaded to an administration system via the monologue link.

WILL TECHNOLOGY CHANGE TREATMENT TECHNIQUES?

When considering new equipment purchases, centres are interested in acquiring MLC and EPI because they offer a potential replacement for conventional lead blocking and portal films. Such use of the MLC will reduce significantly the risks to radiographers associated with the

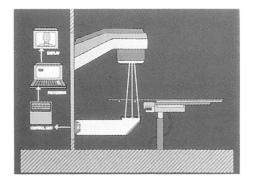

Figure 4.4 Philips SRI 100 Radiotherapy Imaging System.

manual handling of loads. Similarly, EPI will have benefits over film in terms of handling, processing and storage. In the UK it is not currently necessary to have a hard copy treatment verification image for medico-legal purposes but this may become a future requirement and EPI would be a convenient method. However, these technologies also allow radiographers to develop techniques and become more innovative in the delivery of radiotherapy. Development of the radiotherapy service in this way must be considered a key task in any radiographer's role.

Table 4.4 identifies some ideas on how the technology can improve treatment techniques. Some research involving the use of conformal radiotherapy and on line treatment verification has already been carried out on patients undergoing treatment. Dynamic collimation is still in the very early stages of development.

Table 4.4 Improved radiotherapy techniques

Conformal radiotherapy
On line treatment verification
Dynamic collimation

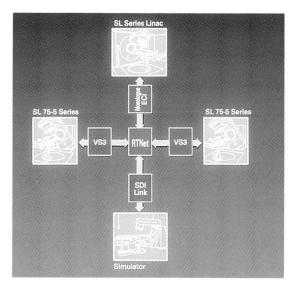

Figure 4.5 Philips Networking System.

Conformal Radiotherapy

By utilizing 3D planning techniques, usually based on CT data, it is possible to design treatment beams shaped to the contours of the tumour and therefore reduce unnecessary dose to normal tissues. This can be achieved with customized blocks but is easier with the MLC. Conformal treatment allows an escalation of dose to be considered which would not be contemplated with conventional volumes.

Horwich et al. (1994) have explored the concepts of conformal radiotherapy and are part of a UK group proposing a dose escalation study in the UK for patients with cancer of the prostate.

The trial would compare conventional dose and fractionation (e.g. 64 Gy in 32 daily fractions), plus or minus a boost (e.g. 10 Gy in 5 daily fractions). The endpoints would be survival, local control, and acute and long-term toxicity.

Dose escalation studies using conformal radiotherapy are already underway in the USA, taking tumour doses beyond 80 Gy.

On Line Treatment Verification

Daily treatment reproducibility and verification are fundamental to the role of radiographers and need input from radiographers at the development stages.

It should be possible to verify field set up before treatment exposure with EPI and make necessary corrections. However, Bissett et al. (1995) have identified that the potential advantage to the patient is currently limited by the operator's ability to verify a treatment image reliably and accurately in real time or to identify any corrective action. In practice, this will only be viable if undertaken by suitably trained radiographers and an automatic process becomes available.

Such a process would provide rapid comparison of the acquired image with a recognized standard, either a simulator verification image or DRR from the 3D planning computer. The comparison would be of field edges, centre and defined structures with a quantifiable measurement of any variation which could be equated with a protocol of allowed tolerances.

Gilderslve et al. (1994) have reported a study of prospective treatment field verification on patients undergoing pelvic radiotherapy. The acquired image was compared to a standard and an assessment made of variation. Although this did not incorporate an automatic process, each assessment was made by the same trained operator. The value of the procedure was demonstrated by the proportion of field placement errors greater than 5 mm being decreased from 69% to 7%.

It would also be useful not only to verify the field reproducibility in this way, but also to verify the dose. The concept of transit dosimetry would use the data acquired to form the image to verify the expected exit dose according to radiological pathlength.

Dynamic Collimation

Dynamic collimation is a technique which will allow precise tissue compensation by moving the machine diaphragms to replace either static wedges or individually customized compensators. Bortfeld et al. (1994) have investigated the use of dynamic collimation in conformal radiotherapy. Such a technique could also be employed in the compensation of fields to improve the dosimetry of tangential breast irradiation. Neal et al. (1994) have identified using the MLC as a means of producing intensity modulation of the beams by utilizing three to six different MLC fields for each tangential beam. They correctly identify that this raises new problems in terms of treatment planning and delivery which must be overcome before such a technique could become available to the vast number of breast cancer patients who would benefit.

WILL TECHNOLOGY INCREASE EFFICIENCY?

Research and Evaluation

Evaluation of new technologies is becoming increasingly important in the NHS (Advisory Group on Health Technology Assessment, 1992). As such it provides an ideal opportunity for radiographers to undertake research studies and to make recommendations about how the various technologies may be utilized effectively.

There are two important examples where such research work has been done in UK, one evaluating the use of a network, and the other comparing the use of the MLC with conventional beam shaping.

Evaluating the Use of a Network

James and Whait (1995) concluded that use of the network provides easy transfer of data and therefore reduces patient waiting times whilst helping to ensure that treatment is efficiently delivered.

However, the report highlighted that patients found sudden transfer to another unit stressful and suggested changes in radiographer procedures to ensure continuity in patient care.

Comparison of the Use of the MLC with Conventional Beam Shaping

Helyer and Heisig, (in press) describe how the techniques of work study were used to compare the time taken for the planning and delivery of shaped radiotherapy beams using a Philips MLC compared to conventional shielding blocks for both simple beam shaping and complex conformal radiotherapy. The results identified time reductions of 19–48% for simple parallel opposed dogleg beams when using the MLC although the time savings were mostly achieved in the daily treatment set up rather than initial planning. For the complex conformal treatments, time reductions of 6–44% were achieved using the MLC with it proving more efficient in both the planning and daily set up. The study also highlighted that more widespread use of the MLC would release the departmental resources employed in the production of customized blocks.

WHAT SKILLS WILL RADIOGRAPHERS NEED?

Radiographers will benefit from improved computing and data management skills, not only because of the computerized nature of the modern linear accelerator, but also to be able to contribute to the development of new systems.

The research base within the profession is beginning to grow and more radiographers are acquiring evaluation and analytical skills as a result. However, these skills need to be applied routinely within the working environment to encourage radiographers to change procedures as new technology comes on line.

Similarly, radiographers need to be proactive in the way they structure their work in order to adapt to, and manage, the inevitable change.

TRAINING ISSUES

Postgraduate education for radiographers is developing rapidly within the higher education sector and there are a substantial number of degree and modular short courses both on line and undergoing validation. Discussion is also beginning within the profession with respect to continuing professional development (CPD) and how this might be achieved. It seems appropriate to consider the training issues identified in Table 4.5 within these two frameworks.

Table 4.5 New technology – training issues

Theory
Practical
Formal/informal
Assessment of competency?

Where?
By whom?

Radiographers will certainly need an understanding of the new technologies and the basic theory could be delivered within an academic setting, but the practical applications will need appropriate clinical input. It seems likely that there will be a need for both formal and informal levels of training and the major manufacturers should perhaps consider addressing some of these needs in addition to their current applications training.

The concept that CPD should be a mandatory requirement for all radiographers is likely to become confused with an argument about who should pay. Equally, however, if left to develop as a professional culture it is also likely to fail unless suitably accredited courses are available with ready access.

The question of the need for competency assessment at a postgraduate level in radiotherapy is extremely controversial and one which the profession needs to debate.

ARE THERE NEW TECHNOLOGIES ON THE HORIZON?

In treatment planning and verification the integration of all imaging modalities will almost certainly come soon, and it would be interesting to consider whether virtual reality techniques could be usefully incorporated into the process.

For treatment delivery the key areas to address must be reproducibility and verification. For example, patient immobilization techniques need an innovative approach and the major manufacturers should become more involved rather than each centre developing local initiatives as tends to be the case currently.

CONCLUSION

Advancing technology as applied to radiotherapy equipment will change the environment in which therapy radiographers will be working. In particular, the linear accelerators currently under development will be very different from those installed only a decade ago. New technology creates an opportunity for the profession to expand roles and take on different responsibilities. Research and development to establish new techniques which may improve patient outcomes but which are both cost-effective and efficient will be within the domain of radiographers. Similarly, radiographer education and training needs will change as will the education provided to other groups within the multi-disciplinary team.

Whatever new technologies become available in the future it is a responsibility incumbent on radiographers to be involved in the design and implementation of the clinical applications arising from those technologies.

Acknowledgements

Grateful thanks are extended to Philips Medical Systems for providing Figures 4.2–4.5.

References

Activity (1994) Pulsed dose rate brachytherapy, radiobiology and initial clinical results. *Nucleotron*, special report No. 5.

Advisory Group on Health Technology Assessment (1992) *Assessing the Effects of Health Technologies, Principles, Practice, Proposals*. London: Department of Health.

Bissett, R., Boyko, S., Leszczynski, K. et al. (1995) Radiotherapy portal verification: an observer study. *British Journal of Radiology* **68:** 165–174.

Bortfeld, T.R., Kohler, D.L., Waldron, T.J. et al. (1994) X-ray field compensation with multileaf collimators. *International Journal of Radiation Oncology Biology and Physics* **28:** 723–730.

Calman, K. (1993) *Hospital Doctors: Training for the Future* PL/CMO(93)3. London: Department of Health.

Expert Advisory Group on Cancer (1995) *A Policy Framework for Commissioning Cancer Services*. London: Department of Health and Welsh Office.

Gilderslev, J., Dearnaley, D.P., Evans, P.M. et al. (1994) A randomized trial of patient repositioning during radiotherapy using a megavoltage imaging system. *Radiotherapy and Oncology* **31(2):** 161–168.

Helyer, S.J., Heisig, S. Multileaf collimation versus conventional shielding blocks: a time and motion study of beam shaping in radiotherapy. *Radiotherapy and Oncology* (in press).

Horwich, A., Wynne, C., Nahum, A. et al. (1994) Conformal radiotherapy at the Royal Marsden Hospital (UK). [Review] *International Journal of Radiation Biology* **65(1):** 117–122.

James, S., Whait, D. (1995) An evaluation of the use of Networking in a busy Radiotherapy Department. *Rad Magazine*, February.

Neal, A.J., Mayles, W.P.M., Yarnold, J.R. (1994) Invited review: tangential breast irradiation – rationale and methods for improving dosimetry. *British Journal of Radiology* **67:** 1149–1154.

Paterson, A. (1994) Developing and expanding practice in radiography. *Radiography Today* 60, No 687.

5
Future Developments in Magnetic Resonance Imaging

Stephen John Blackband

INTRODUCTION

Over the last decade magnetic resonance imaging (MRI) has developed dramatically. Sensitive and difficult imaging techniques have evolved from the confines of the basic research laboratory to the clinical environment. This short chapter attempts to anticipate where clinical MRI is heading over the next few years and is directed at technology and techniques rather than potential clinical benefits. It is expected that given the appropriate technology the clinical utility will follow. This chapter is divided into three main sections. The first section describes the major recent hardware advances from which the near future development of MRI will result. The second section outlines the major recent developments in imaging techniques that will greatly influence future developments, for the most part possible because of the improved hardware outlined in the first section. The final section outlines an optimized MR machine that could be constructed with present technologies. This is not a review, so for the most part only example and general references are included.

HARDWARE DEVELOPMENTS

Gradients

Arguably the most significant development over the last decade has been the introduction of shielded gradients. Early gradient coils, used to spatially encode the signal, were limited in their strength (1 G cm^{-1}) and switching time (0.5–1 ms). Eddy currents, generated by the gradient coils coupling with the magnet, caused significant signal fluctuations often over many milliseconds. This made many sequences, (for example, diffusion, flow and high speed imaging) difficult to implement. In the mid 1980s shielded gradient coils were developed which contrived to eliminate gradient coupling with the magnet (Mansfield and Chapman, 1986; Turner, 1986; Roemer et al., 1986). Gradient systems are now commercially available that can generate 2–3 G cm^{-1} and have switching times of less than 200 µs with minimal eddy currents.

Improved gradients result in a variety of benefits, including shorter echo times, decreased minimum field of view (FOV) and increased number of slices for a given acquisition time. High speed imaging techniques become feasible and sensitive techniques are clinically practical. Gradient switching is also responsible for the loud noise during MR scans that reduces patient tolerance, particularly with rapid imaging sequences. New coil designs that are acoustically screened to greatly reduce this noise (Mansfield et al., 1994) will further improve the acceptability of MR.

Ideally the gradient performance should not be the limiting factor with regard to imaging capabilities, rather it is the signal to noise ratio (SNR) that is the fundamental limiting factor with respect to MRI. Thus once the gradients are improved such that in all situations SNR is depleted before gradient strength, then it will no longer be necessary to improve the gradient strength. This point of view has to be mediated by the possibility of induced nerve stimulation by strong gradients (Mansfield and Harvey, 1995), but it is expected that these effects can be minimized. For example, variants of the acoustically shielded gradient designs reduce neural stimulation (Chapman and Mansfield, 1995).

Phased Array Radiofrequency (rf) Coils

It is well known that the size of the nuclear magnetic resonance (NMR) detector coil is a significant factor in the determination of the

SNR available in the MR experiment; the smaller the coil, the better the SNR. The trade off is a reduction in the FOV of the coil. One approach to reduce this limitation is to construct a large coil out of several smaller coils, each of which receive signal independently so that the image will have the SNR of the smaller coils with a large FOV. The difficulties with such an approach include the need for each coil to have a separate receiver and appropriate circuitry and coil geometry to eliminate coupling between the separate coils (which would otherwise severely distort the images). These coils, designated phased array coils (Roemer et al., 1990), are now commercially available, principally as spine and pelvic coils. However, it is expected that this approach will be extended to other coils to provide additional improvements in other areas of the body. Phased array coils have already made a significant impact in clinical MRI by improving image quality and are expected to have a continuing influence over the next few years.

High Field MRI and Niche Magnets

Since its inception there has been a steady drive to increased magnetic field strengths for MRI and 1.5 Tesla magnets are now clinically commonplace. Going to higher fields may improve the SNR, but it does bring with it several compromising factors, including increased cost, increased radiofrequency (rf) power requirements, leading to increased power deposition in the body, rf coil construction difficulties, increased susceptibility effects and rf penetration problems (Blackband, 1995). Several groups have been evaluating 4 Tesla whole body MR machines and the authors' interpretation is that these associated problems have been difficult to solve. As a compromise, several groups have obtained encouraging results at 3 Tesla and it may prove better to learn difficult lessons on these instruments before moving to higher field strengths.

The gains from going to higher fields in the near future may be better realized on smaller bore instruments dedicated to head and extremity imaging, the so-called 'niche' magnets. This approach may reduce the cost and improve the quality and availability of MRI for specific body regions. For example, based on encouraging high resolution clinical imaging studies of fingers (Blackband et al., 1994), it is anticipated that a small bore dedicated clinical hand imager could be both relatively cheap and effective and have a major impact in orthopaedic medicine and surgery.

Open Magnets

One extremely promising area of MRI under development that may have important consequences on future radiological practices is image guided interventional surgery. Although these methodologies were established using other imaging modalities, MRI brings with it new potential and new difficulties. Standard superconducting magnets constructed as closed cylinders are not easily disposed for surgical access to the patient, and more open magnet designs are required. Siemens have developed a 0.2 Tesla C-shaped electromagnet that allows considerable patient access and reduces patient claustrophobia. General Electric have developed a radically new superconducting magnet design consisting of two parallel rings (Figure 5.1). This 0.5 Tesla magnet has new gradient and rf coil designs to allow open patient access and rapid imaging sequences to be performed. The patient may be inserted along the central bore or through the side of the magnet to allow access for surgical procedures. A range of MR compatible surgical tools have been designed. An ultrasound based tracking system enables the tool position to be monitored and registered on the images. The images may be displayed on a small screen inside the magnet enabling real-time monitoring of interventional procedures. This magnet design also offers other novel possibilities. For example, it is possible to stand in the imaging field so that loaded spines may be examined. This may have significant application in the study of a variety of back problems. Potentially patients could even jog in this magnet! Smaller bore open magnets may also be developed again for niche area imaging, allowing higher fields to be employed for high resolution interventional MRI.

IMAGING DEVELOPMENTS

High Speed Imaging

Once the domain of dedicated research instruments with specially constructed gradient coils, rapid gradient echo-based imaging techniques are now available on several commercial machines. This is largely due to the development of screened gradient coils, the clinical utility of which is presently being explored. In particular, echo planar imaging (EPI) (Mansfield and Morris, 1982) can collect images in 30–40 ms. Improved gradient capabilities have also facilitated

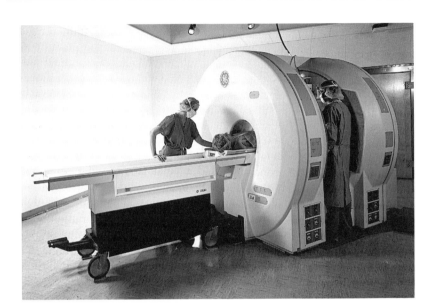

Figure 5.1 Two views of General Electric's 0.5 Tesla open magnet. The 22 inch gap between the two main loops of the magnet allows surgical access.

improvements in high speed spin echo-based imaging sequences. In these methods multiple spin echoes are collected in a single excitation. The gradient capabilities will limit the number of echoes that can be collected in one TR, but 2,4,8 or 16 echoes are now not uncommon on clinical machines in what are termed fast spin echo (FSE) imaging sequences (Listerud et al., 1992). FSE sequences suffer from the disadvantages of a slice selective inversion rf pulse which takes longer to apply than a gradient reversal in a gradient echo sequence (limiting the number of echoes that can be generated during T_2 decay) and increased rf power deposition due to the large number of inversion pulses required. However, FSE sequences have the significant advantage of reducing susceptibility distortions since spin echoes refocus T_2^* effects while gradient echoes do not. This is of great benefit in areas where there are air pockets such as in the abdomen. FSEs have already made a significant clinical impact, especially in body imaging, which is expected to increase over the next few years, especially as the number of echoes obtainable in the echo train increases with improved gradient systems. However, it is the author's opinion that ultimately hybrid spin echo/gradient echo sequences may prove the best way forward, combining the best features of both basic sequences. To utilize these fast imaging techniques effectively a major improvement in clinical imaging consoles is required so that they, too, operate in real time.

High Resolution MRI

MRI is intrinsically SNR limited. Consequently significant increases in spatial resolution can only be achieved if the SNR can be improved. The two main ways this is achieved is by increasing the operating magnetic field strength and/or reducing the size of the NMR detector coil (Mansfield and Morris, 1982). High resolution imaging studies have been performed using small rf coils(1–3 cm) on human skin and fingers at resolutions of 50–100 µm (Bittoun et al., 1990; Blackband et al., 1994). Small diameter or flat 'local' gradient coils are also often employed since standard clinical images do not presently have the necessary gradient strength to achieve these resolutions without significant compromise; however, this situation may change in the near future. Nevertheless high resolution studies have a strong potential clinical utility and may achieve a higher profile with the development of so-called 'niche' magnets of closed or open designs.

Functional Imaging

Decreased gradient eddy currents has also enabled the detection of small (a few per cent) T_2^* generated signal changes in gradient echo images that appear to result from changes in regional function in the human brain, thus termed functional MRI (fMRI) (Ellerman et al., 1994). This technique has great potential for non-invasively examining brain function in a manner similar to positron emission tomography (PET) scanning. Although these effects are most evident at higher fields (since T_2^* effects increase with field strength) they appear detectable at relatively modest fields (1.5 Tesla) providing that rapid imaging techniques are employed to minimize motion artefacts.

It is difficult to predict how far fMRI may develop at this early stage. Certainly considerable money and effort is being invested in its development since the potential rewards are tremendous; the brain is extremely complex and our understanding of it extremely limited. Preliminary animal studies indicate possibilities for fMRI in other body areas. It will be many years, maybe decades, before the full utility of fMRI is realized. Certainly many apparently 'far out' ideas have been proposed. For example, could fMRI form the basis of an intelligence test? The potential of the technique has even reached popular fiction, and in a recent novel (Finder, 1993) the use of a 4 Tesla magnet is accurately described and fMRI is used as a new form of fool-proof lie detector!

Contrast Agents

Three gadolinium based complexes are presently the only commonly used contrast agents in MRI (Runge and Wells, 1995). The heavy metal alters relaxation rates of nearby water molecules, thus altering image contrast. This radiological approach to contrast alteration is extremely useful for lesion detection. However, a range of new contrast agents are under development. In many, gadolinium is complexed to molecular groups that are tissue specific (Runge and Wells, 1995; Proceedings of the SMRM Workshop, 1991) so that particular tissue types and/or lesions may be discriminated. Significant challenges with respect to making the agents safe whilst generating relaxation effects at low concentrations remain. However, if the contrast agents can be made biochemically responsive as preliminary studies suggest, then these agents are expected to have a significant impact in clinical MRI in the near future.

THE MRI MACHINE OF THE FUTURE

It is presently wishful thinking to envisage a 'science fiction' style MR system consisting of, for example, a magnet as large as a room with a complete surgical suite inside, or a very small scanner on the end of a probe that can be waved over the patient (as with Dr McCoy in Star Trek). The following outlines the basis of a machine that could be constructed with present technologies and may be realized in the next decade.

1. A relatively high field (1–2 Tesla) open-style magnet allowing clear patient access. A range of open or closed magnet sizes for niche area imaging would also be available at higher field strengths.
2. Gradients strong enough so that the imaging capabilities are limited by the SNR. A range of interchangeable local gradient coils may be required to practically achieve this particularly a head gradient coil set. Local flat gradient coils are expected to have a role for high resolution imaging. The gradients may also be acoustically screened to improve patient tolerance.
3. A range of rf coils for optimized imaging of all body regions, most of which would be phased array designs and/or flexible.
4. A truly interactive image acquisition console. Images may be collected in real-time and *all* aspects of the imaging sequence are under real-time interactive control (real-time image acquisitions in any orientations can be scrolled through the patient and the scan pulse sequence could be varied in real time). Rather than collect huge amounts of data, a radiographer could scroll through the patient and store data only from regions with pathology.
5. The MR machine should be integratable with a range of surgical devices and other imaging modalities. A range of contrast agents will also play a significant role.
6. Data could be accessed by small hand-held real-time image viewing systems that the radiologist could report from directly and then forward to the referring physician.

CONCLUSIONS

In the limited space for this chapter the major influences in the continuing development in MRI have been outlined. Several other hardware developments will be influential, for example, computer developments, virtual reality technologies, high temperature superconductors, and holography but are considered ancillary to the basic components of the magnet, rf and gradient coils that determine the fundamental abilities of MRI. Also there are a plethora of imaging techniques that will have an increased clinical impact, for example MR angiography, magnetization transfer techniques, diffusion/perfusion imaging, but these are dependent on the basic technologies and enhance MR's capabilities rather than define them. Strictly fMRI fits in this latter category, but has been given special status due to the timing of this chapter and the great potential of the method. NMR spectroscopy has been excluded, since it has yet to be accepted as a clinically applicable technique. This may change in the near future and again developments in clinical spectroscopy have closely followed technical developments. The author believes that a combination of low SNR and patient/tissue heterogeneity will deter the use of spectroscopy as a common clinical technique even though it will remain an important basic research tool.

Acknowledgements

Many thanks to Dr Derek Shaw for supplying the figure on behalf of General Electric, to Dr Peter Gibbs for critiquing the manuscript and to Dr Richard Bowtell for some useful comments.

References

Bittoun, J.B., Saint-James, H., Querleux, B.G. et al. (1990) In vivo high resolution MR imaging of the skin in a whole body system at 1.5 T. *Radiology* **176**: 457–460.

Blackband, S.J. (1995) Introduction to the physical principles of NMR and MRI. In Tempany, C. (ed.) *MR and Imaging of the Female Pelvis*. St Louis: Mosby Year Book, pp. 1–34.

Blackband, S.J., Chakrabarti, I., Gibbs, P. et al. (1994) Fingers: three-dimensional MR imaging and angiography with a local gradient coil. *Radiology* **190**: 895–899.

Chapman, B.L.W., Mansfield, P. (1995) A quiet gradient-coil set employing optimised, force shielded, distributed coil designs. *Journal of Magnetic Resonance B* **107(2)**: 152–157.

Ellerman, J., Garwood, M., Hendrich, K. et al. (1994) Functional imaging of the brain by nuclear magnetic resonance. In Gilles, R.J. (ed.) *NMR In Physiology and Biomedicine*, pp. 137–150. San Diego: Academic Press.

Finder, J. (1993) *Extraordinary Powers*. London: Orion Books.

Listerud, J., Einstein, S., Outwater, E., Kressel, H.Y. (1992) First principles of fast spin echo. *Magnetic Resonance Quarterly* **8(4)**: 199–244.

Mansfield, P., Chapman, B. (1986) Active magnetic screening of coils for static and time dependant magnetic field generation in NMR imaging. *Journal of Physics E* **19**: 540–545.

Mansfield, P., Harvey, P.R. (1995) Limits to neural stimulation in echo planar imaging. *Magnetic Resonance Medicine* **29(6)**: 746–758.

Mansfield, P., Morris, P.G. (1982) NMR imaging in biomedicine. *Advances in Magnetic Resonance Suppl 2*. New York: Academic Press.

Mansfield, P., Bowtell, R., Glover, P. (1994) Active acoustic screening – design principles for quiet gradient coils in MRI. *Measurement Science and Technology* **5(8)**: 1021–1025.

Proceedings of the SMRM Workshop on Contrast Enhanced Magnetic Resonance (1991) *Magnetic Resonance Medicine* **22(2)**: 177–378.

Roemer, P.B., Edelstein, W.A., Hickey, J.S. (1986) Self shielded gradient coils. *Proceedings of the 5th Annual Meeting Society of Magnetic Resonance in Medicine (Montreal)*, August, Vol 3, pp. 1067–1068.

Roemer, P.B., Edelstein, W.A., Hayes, C.E. et al. (1990) The NMR phased array. *Magnetic Resonance Medicine* **16(2)**: 192–225.

Runge, V.M., Wells, J.W. (1995) Update: safety, new applications, new MR agents. *Topics in MRI* **7(3)**: 181–195.

Turner, R. (1986) A target field approach to optimal coil design. *Journal of Physics D* **19**: L147–L151.

6

Moral Dilemmas in Diagnosing Fetal Anomalies Using Ultrasound

Regina Fernando

INTRODUCTION

Over the last decade technological improvements in ultrasound equipment have led to enhanced fetal imaging. An increasing number of fetal malformations can now be detected. These improvements have had a profound influence on the debate over medical ethics. It takes time for ethical principles to evolve. A vast amount of thought and discussion always precedes the construction of moral guidelines for society. It is beyond the scope of this chapter to enter into a discussion on the exact nature of morals and ethics. Morals will be addressed in a general sense. In particular some of the moral dilemmas that sonographers are exposed to due to the use of ultrasound in the diagnosis of fetal anomalies will be considered. These include the questions of whether to encourage routine scanning and whether informed consent for a scan should be sought, together with the controversies on 'soft markers' and elimination of defective fetuses.

MORALS AND MORAL DILEMMAS

Morals are concerned with the rights and wrongs of everyday life. According to Beauchamp and Childress (1994) the term morality 'refers to social conventions about right and wrong human conduct that are so widely shared that they form a stable (although usually incomplete) communal consensus'. This definition intimates that views of right and wrong are not complete. They can change with human experience. The human experience of diagnosing fetal abnormalities prenatally is still very much in its infancy. Moral dilemmas occur when there is a state of indecision between two alternatives. Beauchamp and Childress (1994) describe two forms of dilemma. The first is where there is some evidence to indicate that an act is morally right and some evidence to indicate that it is morally wrong, but the evidence on both sides is inconclusive. Abortion could be considered such a dilemma for those who see the evidence in this way. The second form is where there are two morally correct ways of acting, x and y, in a situation. However, doing x precludes doing y. The evidence supporting both x and y is good, but neither set of evidence is dominant. Moral dilemmas in health care often arise due to the complex interaction within institutional and societal structures (Purtilo, 1993). Many of the sonographer's moral conflicts arise because he/she must respond to demands imposed by government policy and consumer groups as well as to the demands of his/her own and other health care professions.

THE 18–20 WEEK ROUTINE SCAN

In the United Kingdom (UK) all pregnant women are routinely offered an ultrasound scan in the second trimester (18–20 weeks). This offers the greatest potential of picking up fetal anomalies (RCOG, 1991). Research in the UK supports this as a beneficial practice (Chitty, 1994; Chitty et al., 1991; Shirley et al., 1992). However, there is evidence to suggest that the use of ultrasound pre-natally does not improve perinatal outcome (Bucher and Schmidt, 1993; Ewigman et al., 1993). Furthermore, questions have been raised about the safety of ultrasound which has not been validated by large scale clinical trials (Newham et al., 1993; Salveson et al., 1993). Therefore, should sonographers encourage the performance of the 18–20 week 'routine' screening anomaly scan? Why scan if there is an equivocal outcome and a possibility of harm? (Figure 6.1) Routine scanning would appear to contradict two major ethical principles of health care, the principle of non-maleficence ('do no harm') in cases where there is the potential to do harm to the fetus, and beneficence ('do good') where the doing of good cannot be proved. However, respect for the patient's autonomy, or

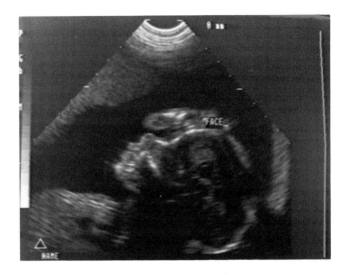

Figure 6.1 Normal scan – non-maleficence or beneficence?

the right to self-determination, is another central principle of medical ethics. It is the mother's right to expect to partake of the health care options available throughout her pregnancy. The anomaly scan may be considered by the mother for a multitude of reasons. Some of these reasons have been demonstrated by Berwick and Weinstein (1985) who found that less than 50% of a mother's willingness to pay for an ultrasound scan was for information on the normality of the fetus. More than one-third of the value attached to the scan was for information that could not possibly affect her clinical management.

INFORMED CONSENT

In the UK the debate on informed consent in the decision to have an 18–20 week ultrasound scan is still ongoing. The present position is that the woman must 'opt out' of anomaly screening. Gaining informed consent from the patient for most routine antenatal screening procedures rarely occurs (Jones, 1994). The less invasive the procedure, and ultrasound is considered one of the least invasive, the less likely that prior counselling will have occurred. Many antenatal departments distribute written information about the test but few offer verbal information or check to see if the written information has been read and understood. If the woman does not ask any questions it is assumed that she understands and wants the test (Jones, 1994). When a woman presents for her anomaly scan it is often obvious that understanding of the reasons for the scan and its possible consequences is limited. Should the sonographer inform the woman of the potential outcome and check on her knowledge, i.e. seek informed consent or should he/she ignore these problems? The problems relating to non-maleficence and beneficence are again at play here. Full informed consent for the scan is required to satisfy the duty of non-maleficence, but not providing information on the potential outcome may increase beneficence as anxiety will not be raised.

It is the duty of anybody carrying out a test on a patient to ensure that the patient is fully informed and consents to the test (Purtilo, 1993). Chervenak and McCullough (1994) outline several stages in the procedure of informed choice for an ultrasound examination. First, the woman should be provided with information about the actual and theoretical benefits and limitations of obstetric ultrasound. She should then evaluate this information in terms of her own values and articulate her preferences. The obstetrician should then provide the woman with his/her own recommendation for that particular pregnancy. Any disagreement should be sensitively discussed before the woman makes her final decision. This method has resource consequences in terms of available time and may be more suitable for the American health system from where it originates than for the UK system. Nevertheless, the authors argue that in Western democracies no credible argument based on excessive cost has as yet been advanced to outweigh the consideration of autonomy.

'SOFT MARKERS'

The increased incidence of soft markers or anatomical variants which may indicate an increased risk of chromosomal abnormality has exacerbated sonographic dilemmas. The prevalent soft markers are choroid plexus cysts, dilated renal pelves and echogenic foci in the ventricles of the heart. The soft marker which is most likely to cause anxiety is the choroid plexus cyst as it is a developmental variant in the cerebral anatomy, i.e. the brain tissue. The first report of choroid plexus cysts appeared in the literature over a decade ago (Chudleigh et al., 1984) but there is still no consensus as to their management. For obvious reasons any suggestion of a deviation from normal in the appearances of the brain raises anxiety in many parents (Figure 6.2).

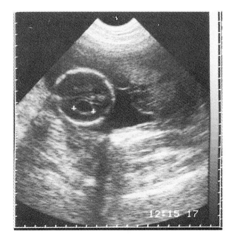

Figure 6.2 Choroid plexus cyst – a normal variant or an abnormality marker?

The following hypothetical case illustrates the sort of dilemma that arises.

Jane has recently started work in hospital A, having been employed for the previous three years as a sonographer at hospital B. Jane has performed a 'routine' 19 week anomaly scan and detected a choroid plexus cyst. No other abnormalities or variants are detected. At her previous hospital all patients with choroid plexus cysts were counselled by an obstetrician and offered amniocentesis to exclude chromosomal anomalies. At this hospital amniocentesis is not offered and the sonographers are expected to explain the findings and reassure the patient. Jane feels that she cannot explain the findings without suggesting the additional risk of a chromosomal disorder. Should she withhold the information on possible chromosomal links and concentrate on the positive aspects such as the fact that choroid plexus cysts resolve spontaneously at approximately 24 weeks and have no neurological sequelae; or should she disclose the link between choroid plexus cysts and chromosomal problems and risk raising maternal anxiety for the rest of the pregnancy?

Withholding information is paternalistic and is a commonly confronted problem in health care. It is often confused with caring but in fact disregards the patient's autonomy. In reality it is used as a method of avoiding awkward situations and perhaps further discussion and is more likely to benefit the health worker than the patient.

THE PATIENT VERSUS SOCIETY

The 18–20 week anomaly scan has been promoted with a view to eliminating fetuses with some defect or abnormality. The assumption is that the mother will want to terminate the pregnancy if a serious fetal disorder is found. Sonographers with strong moral or religious objections to abortion may have difficulty promoting such a practice. Nevertheless, the anomaly scan does have therapeutic potential as well. It relieves the mother of a previously abnormal pregnancy of months of anxiety; it allows for successful interventional procedures to take place in certain circumstances, for example, drainage of fetal ascites and thoracic effusions, thus allowing a healthy child to be delivered at term, and it alerts the physician to possible difficulties in delivering an abnormal fetus.

There is a prevailing attitude that it is better for the mother, for society and for the affected child itself, if it is not born (Lawson, 1995; Sutton, 1990). This attitude has also been evident in work by Schuth et al. (1994). This group reported that 17 out of 29 women who continued with a pregnancy after fetal malformation was diagnosed by ultrasound felt that the reactions of medical staff were negative and hurtful. These mothers felt pressured to justify themselves rather than gaining respect and encouragement for their decisions. It is possible, too, that selective termination may well undermine the position of the disabled and promote a more intolerant society.

COMMUNICATION

When a fetal malformation is diagnosed by ultrasound, communication of results and the subsequent counselling of patients and their partners can be a source of anxiety for all parties involved. The sonographer who initially detects the abnormality may be working under restrictive local rules which do not allow disclosure of information, or allow the delivery of limited information only. If the patient continues to question the sonographer there may be no alternative but to describe fully the scan appearances which may put him/her into a compromised position. The dilemma here is to tell the truth and risk the harm that could result from unwisely interfering in the doctor/patient relationship or to say nothing and deprive the patient of adequate care in the situation. In practice something is normally said to comfort the patient and explain the sonographer's dilemma to the patient. These situations are, happily, decreasing in numbers due to increasing awareness of professional rights amongst sonographers and recognition of sonographers' expertise (College of Radiographers (COR), 1994; Saxton, 1992).

It must also be noted that whatever is said in cases where there are anomalies may be misconstrued. This may be due to the psychological state of the patient or the communication skills of the sonographer, or a combination of both. It is not always possible for sonographers to be as sensitive as they would like or to assess accurately the psychological needs of each individual patient. Sonographers are often blamed for inadequate psychological care and counselling immediately following the detection of fetal abnormalities. This has been noted in other professions in similar circumstances. In a study on parents' needs after ultrasound diagnosis of a fetal malformation it was suggested that the doctor is often used as a scapegoat in order to assuage feelings of self reproach and guilt (Schuth et al., 1994). Responsibility can be abdicated to the doctor particularly with regard to the decision to terminate. It was reported that when some time had elapsed after the event, patients believed that the doctor was a central contributing factor in the decision to terminate the pregnancy, although objectively this may not have been the case.

CONSCIENTIOUS OBJECTORS

The Abortion (Amendment) Act of 1990 allows termination of a seriously mentally or physically handicapped fetus up to and including the expected date of delivery. Some abnormalities may not be detected until late in the pregnancy. This may be due to the patient presenting late for booking or the failure to detect the anomaly earlier, for example as in some cases of cardiac anomalies. In the case of multiple pregnancies where one fetus is affected by a malformation selective termination of the affected fetus may be considered. Late termination and selective fetocide involve a lethal injection administered to the fetus in utero. This injection is normally performed under ultrasound control and sonographers may be asked to assist.

This raises the issue of conscientious objection which has been considered in the current Code of Professional Conduct for Radiographers (COR, 1994). Radiographers should 'make known' any conscientious objection which may be relevant to their clinical practice. However, the

Code of Conduct does not consider how this might be done. In view of current sonographic practices more detailed guidelines are urgently required to support sonographers as independent practitioners. Midwives, widely recognized as independent practitioners (Jones, 1994), have many years experience in the area of abortion and it seems appropriate for radiographers to take their cue from them in considering and developing more detailed guidelines. Those midwives with strong ethical objections put them in writing to their departmental managers; they may then be excused from partaking in the actual procedure but not from pre and post procedure care of the patient (Jones, 1994). However, the Abortion Act points out that in the case of an emergency occurring a midwife would be expected to carry out whatever care is required during the procedure. The circumstances and working relationships in obstetric ultrasound are complex. There is a requirement to outline what might constitute pre and post procedure care in obstetric sonography in order to avoid potentially ambiguous situations.

Sonographers must make a judgement and act in a way that protects their own moral integrity. Beauchamp and Childress (1994) indicate that individuals ought to do everything possible to ensure that relevant moral principles and rules inform their consciences. Religious convictions aside, the moral principles and rules surrounding termination are debatable and of great concern to society. Sonographers have a duty to care for their patients but they should not be expected to compromise themselves.

CONCLUSION

Sonographers are confronting moral issues that have not previously been considered. Ethical debate is developing on the value of routine scanning, informed consent for ultrasound scans and the management of 'soft markers'. Conscientious objection is currently a low key issue in sonography. However, late termination and selective fetocide are gaining acceptance and the ongoing developments in fetal medicine are likely to highlight this issue in the very near future.

The current debate emanates predominantly from the medical profession and it could be argued that sonographers are passive. As the keyholders of the obstetric ultrasound service (Scott-Angell and Chalmers, 1992) sonographers ought to be leading this debate. At the very least a strong representation of sonographic interest in these issues provides balance. There is a need to recognize the rights of the fetus, mother, sonographer, health care service and society. Sonographers must actively participate in the debate if their rights are to be adequately considered. There are no easy solutions to many of the dilemmas which arise daily but opening them up for public debate will no doubt help in the formation of ethical guidelines for the future.

COMMENT Audrey M. Paterson, *Canterbury*

It is essential, of course, that sonographers participate actively in the moral and ethical debates that surround their practice in obstetrics. Unhappily, there is little evidence to suggest that sonographers are engaging in such debates and almost no evidence that they are leading them. This despite twenty years, or more, as key holders of the service.

Throughout the history of obstetric ultrasound, ethical and moral concerns have been expressed, but largely by groups other than sonographers. For example, Winkler and Goodwin (1988) reported on a survey on access of companions to obstetric ultrasound departments. The survey was carried out for the Greater London Association of Community Health Councils following reports from women on the treatment of their partners and families during obstetric scans. While Winkler and Goodwin found the results of their survey to be encouraging, it was the case that almost a third of the hospitals surveyed had policies that restricted access to some degree. Today, almost a decade later, and following promotion of the Patient's Charter, it would be reasonable to expect that this ethical issue would be dead; to expect that all sonographers would allow the women using their services the right to determine whether and when partners and families should participate in the scan.

Unfortunately, as McInnes (1995) reported, this is not yet the case and 'many radiographers (sonographers) and ultrasound departments all around the country [who] still feel strongly about this issue'. If sonographers are unable to resolve this ethical debate, such that unacceptable practice still continues in some parts of the United Kingdom, what hope is there for their role in the deeper and more complex dilemmas associated with their practice?

Recently, and quite independently, two publications have raised serious ethical issues for sonographers, but, once again, the voice of the sonographer on the matters raised has barely been heard. In 1993, the Association for Improvements in the Maternity Services (AIMS) produced a special edition of its journal entitled *Ultrasound ??? Unsound* (Lawrence Beech and Robinson, 1993). Within it, they expressed the concerns they had been raising with Ministers of Health for twelve years about the widespread, effectively routine, use of ultrasound unsupported by proper studies to evaluate its effectiveness and potential risks. In 1994, the second publication, the report of the Independent Inquiry into Obstetric Ultrasound Procedures at the University Hospital of Wales was produced (Case et al., 1994). This report of a public inquiry was conducted following a misdiagnosis in which a woman was told that her pregnancy was no longer viable and that her baby was dead. A similar misdiagnosis had been made during this woman's first pregnancy, three years earlier. Both misdiagnoses were made following ultrasound scans.

The contents of both of these publications demand that sonographers engage in significant, organized and public debate about their practice and its ethical and moral consequences. Sonographers must participate in, and lead, work that investigates properly the effectiveness of ultrasound as well as its risks. The need is ever more urgent as the imaging capabilities of ultrasound machines continue to improve and the number of soft markers and anatomical variants that it is possible to demonstrate continues to grow. Equally, sonographers must lead the questioning of the appropriateness of some scans, and of the adequacy of some scan operators. In relation to the University Hospital of Wales Inquiry it could be suggested that a dubious referral (for 'peace of mind' – but whose is the question?) for the scan and an insufficiently trained and supervised practitioner contributed significantly to the misdiagnosis.

As has been stated earlier, sonographers are the key holders of the obstetric ultrasound service. The use of ultrasound imaging in obstetric practice is now so widespread and its ability in trained hands to detect fetal anomalies is now so good that at least one department works 'on the premise that all fetuses have an abnormality unless proved otherwise' (Walton et al., 1995). The potential for harm to normal fetuses, to pregnant women and their families, and to the wider society from such a working premise is unbounded – and is unacceptable and unsustainable without a full and continuing ethical and moral debate to accompany it. Sadly, amongst sonographers, there is little to demonstrate that such a debate exists.

References

Beauchamp, T.L., Childress, J.F. (1994) *Principles of Biomedical Ethics*. Oxford: Oxford University Press.

Berwick, D.M., Weinstein, M.C. (1985) What do patients value? Willingness to pay for ultrasound in normal pregnancy. *Medical Care* **23**: 881–893.

Bucher, H.C., Schmidt, J.G. (1993) Does routine ultrasound scanning improve outcome in pregnancy? Meta-analysis of various outcome measures. *British Medical Journal* **307**: 13–17.

Case, J., Campbell, S., Hately, W. (1994) *Independent Inquiry into Obstetric Ultrasound Procedures at the University Hospital of Wales, Report – April 1994*. Cardiff: South Glamorgan Health Authority.

Chervenak, F.A., McCullough, L.B. (1994) Should all pregnant women have an ultrasound examination? *Ultrasound Obstetrics and Gynecology* **4**: 177–180.

Chitty, L.S. (1994) Routine fetal anomaly scanning: the case in favour. *BMUS Bulletin* **2(1)**: 31–32.

Chitty, L.S., Hunt, G.H., Moore, J., Lobb, M.O. (1991) Effectiveness of routine ultrasonography in detecting fetal structural abnormalities in a low risk population. *British Medical Journal* **303**: 1165–1169.

Chudleigh, P., Pearce, J.M., Campbell, S. (1984) The prenatal diagnosis of transient cysts of the fetal choroid plexus. *Prenatal Diagnosis* **4**: 135–137.

COR (1994) *Code of Professional Conduct*. London: The College of Radiographers.

Ewigman, B.G., Crane, J.P., Frigoletto, F.D. et al. (1993) Effect of prenatal ultrasound screening on peri-natal outcome. *New England Journal of Medicine* **329**: 821–827.

HMSO (1990) *Abortion (Amendment) Act 1990*. London: HMSO.

Jones, S.R. (1994) *Ethics in Midwifery*. London: Mosby Year Book Europe.

Lawrence Beech, B., Robinson, J. (1993) Ultrasound unsound *AIMS Journal* **5(1)**.

Lawson, D. (1995) All you need is life. *The Spectator*, 17 June, 15–16.

McInnes, E. (1995) Ultrasound update – the continuing debate! *Synergy*, **September,** 49.

Newham, J.P., Evans, S.F., Michael, C.A. et al. (1993) Effects of frequent ultrasound during pregnancy: a randomised controlled trial. *Lancet* **342**: 887–891.

Purtilo, R. (1993) *Ethical Dimensions in the Health Professions*. Philadelphia: W. B. Saunders.

Royal College of Obstetricians and Gynaecologists (1991) *Report of Study Group on Antenatal Diagnosis of Fetal Abnormalities*. London: RCOG.

Salveson, K.A., Vatten, L.J., Eik-Nes, S.H. et al. (1993) Routine ultrasonography in utero and subsequent handedness and neurological developments. *British Medical Journal* **307**: 159–164.

Saxton, H.M. (1992) Editorial, Should radiologists report on every film? *Clinical Radiology* **45**: 1–3.

Schuth, W., Karck, U., Wilhelm, C., Reisch, S. (1994) Parents' needs after ultrasound diagnosis of a fetal malformation: an empirical deficit analysis. *Ultrasound Obstetrics and Gynecology* **4**: 124–129.

Scott-Angell, N., Chalmers, R.J. (1992) Results of a survey on ultrasound examinations. *Radiography Today* **61**: 34–37.

Shirley, I.M., Bottomley, F., Robinson, V.P. (1992) Routine radiographer screening for fetal abnormalities by ultrasound in an unselected low risk population. *British Journal of Radiology* **65:** 564–569.

Sutton, A. (1990) *Prenatal Diagnosis: Confronting the Ethical Issues.* London: Linacre Centre.

Walton, B. et al. (1995) Setting the Record Straight *Synergy*, **November,** 16–17.

Winkler, F., Goodwin, J. (1988) *Access of Companions to Obstetric Ultrasound Departments, Report of a Survey for the Greater London Association of Community Health Councils.* London: Greater London Association of Community Health Councils.

7
The Nature of Image Reporting

Philip J. A. Robinson

INTRODUCTION

Amongst the current controversies in medical imaging are the questions of whether reports on imaging procedures are always necessary, who should carry out reporting, and whether reporting should form an integral part of an imaging procedure or be regarded as a separate activity. This chapter will not attempt to address any of these issues but will focus on an aspect of reporting which has been largely ignored so far – the question of what actually constitutes a 'report'.

WHAT IS A REPORT?

In attempting to define a report in the context of diagnostic imaging it may be useful to describe the content and properties of reports as determined from analysis of a large sample, but first it is important to illustrate what a report is not.

A Report is a Result, but a Result is not a Report

Whereas the results of biochemical tests or physiological measurements are typically expressed in terms of a numeric variable which can be compared against an established normal population range, the results of imaging procedures always involve the additional step of interpretation of the image, even when a measured variable forms a major part of the report. For example, measurement of the fetal biparietal diameter may be the single most important component of an ultrasound report but obtaining such a measurement still requires a considerable input of technique and skill from the operator. The measurement of individual renal function on a radionuclide study may be expressed in numerical terms in the report, but the derivation of the figures still requires an interactive process depending upon visual interpretation of the images by the operator.

A Report may Include a Description, but a Description is not a Report

Descriptive elements in imaging reports have two functions. The first is very similar to the situation where a spectator watching a cricket match may simultaneously listen to the radio commentary in case the commentators, who are allegedly experts in the subject, may spot something that the spectator missed. This is clearly much more likely if the spectator is unfamiliar with the game but even the most experienced observer might well overlook some details. The second objective of a description in image reporting is to act as a verbal back-up to the images themselves, primarily for the benefit of those subsequent readers who do not have the images in front of them. This is most likely to be the case for the review of historic reports, when images have been destroyed, and for reports to general practitioners or others who do not routinely receive the films.

A Report is, or Includes, an Opinion

In addition to descriptive elements, reports typically contain interpretative comments indicating the significance of the visual findings in terms of the clinical problem posed by the patient. It is helpful to bear in mind that this type of interpretation may be properly called an 'opinion', and that all opinions, by definition, are based on a degree of ignorance – at least to the extent that where matters of fact or convention are established, opinions are superfluous. For example, there is no point in having an opinion about what day of the week follows Tuesday or how many miles it is between Leeds and London. An opinion, then, is a speculation based on past experience and current observation.

THE MEANING OF 'FINDINGS'

The descriptive and interpretational elements of a report can usefully be combined under the general term 'findings'. The production of a report of 'findings' involves three steps – perception, interpretation and diagnosis. Perception is a matter of answering the question 'what do the images show?' This requires visual recognition of the image features and mental comparison with historically recollected data describing normal and abnormal appearances. The phase of interpretation involves answering the question 'what produced these appearances?' This requires an understanding of the mechanisms of disease or trauma which cause abnormal appearances, and also an understanding of the range of normal variation and changes with age. The phase of diagnosis requires an answer to the question 'what do these appearances mean in terms of pathology?' In other words, can the observer deduce from the appearances the nature, extent and severity of the underlying disease process? In the interests of clarity and brevity, the thought processes described above are not usually encapsulated in the report but are taken as read and, in most cases, the report author can go straight to a diagnostic conclusion without recording the intervening logical steps. For example, a report such as 'there is a sub-capital fracture of the left femur' makes a whole series of assumptions regarding the derangement of local anatomy together with a mechanism for injury, but such a report may be entirely justified from a knowledge base of past experience.

A further element of some reports involves recommendations for future steps in the management of the patient. These might be suggestions for additional imaging procedures in case of persisting doubt about the diagnosis, or for follow-up imaging, but may also include proposals for interventional therapy. Because this element of reporting has different implications from the diagnostic elements so far described, it is probably better to consider it separately (see below).

WHAT CONCEPTS DO REPORTS CONTAIN?

Robinson and Fletcher (1994) recently analysed a sample of 6400 text reports obtained from two large radiology departments. Included were reports of plain films, contrast procedures, ultrasound, computed tomography (CT) and nuclear medicine examinations (Table 7.1). The terminology used was extracted and classified into generic groups,

Table 7.1 Reports sampled for analysis

Examination type	Number in sample
Skull/spine	500
Upper limb	500
Lower limb	500
Barium meal	500
Barium enema	500
IVU	500
Ultrasound	1000
CT head	500
CT body	500
Bone scans	400
Random sample of all examinations	1000
Total	6400

using as a starting point the two major headings from the American College of Radiology (1992) diagnostic classification – anatomy and pathology. Subsets of 'pathology' terms were developed, including diagnoses, conditions and observations (described below). The terms found in the sample also included details of technique, adverse events or other limitations of the procedure, and terms relating to medical or surgical treatment (for example, comments on prostheses, catheters, clips etc.). The residue of the report sample was made up of concepts describing the context of the procedure and its interpretation. These could be grouped under three main headings – qualifiers for severity or degree, qualifiers for probability or uncertainty, and qualifiers for progression of disease with time or in relation to treatment. Table 7.2

Table 7.2 Concepts contained in reports

Anatomic terms
Details of examination technique
Limitations and adverse events
Findings
   pathologies
   diagnoses
   conditions
   observations
Qualifiers for severity, degree, and extent
Qualifiers for probability or uncertainty
Qualifiers for time-scales and progression
Recommendations for future management

summarizes the classification of concepts used in reporting. Qualifiers for severity included descriptive terms extending from the distinction of normality from abnormality, through estimates of extent, number, size and grading, to general descriptions of severity and degree. Qualifiers for probability or uncertainty included comments relating to significance, evidence, exclusion and correlation with clinical findings. Qualifiers for progression were used in two senses. Firstly as an indicator of the progression of disease between consecutive examinations (for example, improved, worse, no change) and, secondly, on a single examination where comment was made on the likely duration of the disease process shown (examples included acute, chronic, recent, healed, long-standing, active).

The categories for classifying the 'pathology' terms were as follows, in decreasing order of specificity:

1. True pathology terms, including specific pathologies, for example, parathyroid adenoma; and also general pathology terms such as ischaemia, infarct, and neoplasm.
2. Diagnoses – these included entities which were specific enough to be valid as clinical 'diagnoses' but which may be caused by different underlying pathology, for example, hyperparathyroidism.
3. Conditions – this arbitrary term described those entities whose radiologic appearances were specific enough to show the site and type of lesion present but were non-specific for pathology. An expanding bone lesion in a rib and a local stricture in the colon are two such examples.
4. Observations – these were features which were recognizably abnormal on images but which have no specific equivalent in pathology. Examples include an opacity on a chest X-ray, reduced attenuation on a CT image, and increased echogenicity on an ultrasound image.

These categories are to some degree artificial and they do overlap. Their main value is to allow correlation of groups of concepts with existing classifications of terms. A fuller account of this analytical work has been presented elsewhere.

A MODEL FOR THE STRUCTURE OF DIAGNOSTIC IMAGING REPORTS

The analysis summarized above suggests that it should be possible to derive a generic model for the structure of imaging reports. The model assumes that the report can be described as a series of data elements which answer the following questions:

1. What exactly was done? – a description of the technique and its variations.
2. Did anything go wrong? – a description of the limitations of the procedure and any adverse events which took place.
3. What did the procedure show? – an account of the findings.
4. What should be done next? – recommendations for subsequent imaging or therapeutic intervention.

In many cases, particularly plain film examinations and those procedures with normal results, an abbreviated or coded comment on findings may be all that is needed.

WHAT IS A 'GOOD' REPORT?

Strictly speaking, the measurement of quality is impossible. By definition, 'quality' describes those attributes of an entity which are unmeasurable (if they were measurable, they would be 'quantity'). Quality, like beauty, is in the eye of the beholder; it can only be measured indirectly, by selecting attributes which are generally agreed to contribute to 'quality' and which themselves are measurable. This is not an easy task – hence the rather limited appeal of so-called quality indicators for radiology such as waiting times, recall rates, and cancellation rates (Royal College of Radiologists, 1995). At the time of writing, the only indicators for measuring the quality of image reports which have so far been suggested relate not to the content of reports but to the process of reporting, with emphasis on the speed of production of reports and their ready availability at the site of application (Royal College of Radiologists, 1995). Attempts to investigate the validity of the content of reports form a legitimate part of clinical audit in radiology, but sensible audit studies of image reporting are very difficult because of the lack of agreed standards against which performance can be measured.

The medico-legal position in the United Kingdom simplifies these considerations by defining acceptable professional performance as that level of performance which would be expected to be achieved by the majority of practitioners having the same experience and responsibilities as the individual being scrutinized. Broadly speaking, what this means is that everyone should get the easy cases right (that is what the easy cases are – those which everyone should get right), whereas it would be quite legitimate to make a few mistakes with 'difficult' cases. Professional performance is not expected to be perfect and it is accepted as normal that occasional errors will be made. Whether a particular error is acceptable or not revolves around the opinion of experts in the specialty. This approach is related to a technique used in radiology research where a 'gold standard' report may be produced by consensus of a group of experts. The hypothesis is that the consensus opinion can then be regarded as the 'correct' report, and reports from all other observers can be compared against this benchmark.

Errors and omissions in reporting are traditionally classified into errors of perception and errors of interpretation (Kundel, 1989). It has been clearly shown that both types of error can be reduced by double reporting either using two different observers or with the same observer reporting the images twice (Yerushalmy et al., 1950; Markus et al., 1990; Anderson et al., 1994). These research studies have focused largely on test sets of images about which specific questions are asked, for example whether radiographs of the lumbar spine are normal or abnormal, whether a lung scan shows evidence of pulmonary embolism or not (Deyo et al., 1985; Gjorup et al., 1986; Gray et al., 1984). The general conclusions which can be drawn from such studies are:

1. The performance of all observers is to some degree erratic.
2. Observer variation in detecting gross abnormalities is much less than variation in detecting subtle abnormalities.
3. Errors of perception can be reduced by training – the observer can learn what to look for.
4. Errors of interpretation also become less frequent with greater experience.

Studies have shown that when considering test cases with binary type reporting decisions (fracture versus no fracture; presence or absence of a lung nodule), the improvement in performance which comes with increasing experience is largely due to better specificity rather than better sensitivity (Sheft et al., 1970; Pauli et al., 1996). The suggested explanation for this is that all biological observations have a variable range of normal and the more experienced observers will have seen a greater number (though not a greater proportion) of 'extreme' examples of the normal range, as well as having a greater experience of abnormal cases. This finding reflects a general feature of medicine and biology – that becoming educated is largely a matter of learning to recognize what is not true.

UNDERSTANDING THE QUESTION

If it is agreed that a report is a written or verbal commentary accompanying a set of images, what attributes should be put into it? The answer to this question would be immediately obvious once the question *'what is the report for?'* is understood. Attempts to define general objectives for diagnostic image reports will fail because the aims of imaging procedures vary both in terms of the enormous range of possible pathologies and also in the variety of clinical contexts in which imaging procedures are requested. For example, a postoperative chest radiograph may be requested to look for evidence of lobar collapse, for signs of a pneumothorax, or for demonstration of the position of a central venous catheter. Not only might the operator employ a different radiographic technique in each of these cases, but the report itself will focus on different observations.

Reporting images is an exercise in communication. In order to succeed, the reporter needs to send the right message at the right time to the right person. Correct interpretation of the images is only a part of the more difficult problem of understanding what message to send. It may be argued that image interpretation is a secondary requirement in reporting – the primary requirement is for the reporter to understand why the examination was done. Once the reporter knows what clinical questions are being asked, and what their context is in the management of the individual patient, then the image interpretation becomes much more focused (though not always easier). In many cases both the clinical question and the context are very straightforward. For example, a casualty referral for radiography of the wrist with the clinical information 'fell on outstretched hand' is easy to understand. However, a referral for chest radiography with the clinical information 'chest pain? cause' requires in the reporting author several sets of knowledge. Firstly, the author must be aware of the range of conditions causing chest pain and their possible manifestations on the radiograph. An operator who is familiar with this clinical knowledge should be able to report chest radiographs with a high level of sensitivity, that is, the operator should not miss any abnormalities. Secondly, however, in order to report with a high level of specificity, the operator also needs to know how often all the features found on chest radiographs occur in association with the particular diseases under question, and how often they occur in patients without these diseases. Finally, in order to interpret the significance of the radiographic observations, the reporter would need to know the frequency of chest pain as a manifestation of each of the conditions under consideration, and further, a knowledge of the prevalence of each of these conditions in the population to which the patient belonged. 'Understanding the question' then becomes a very difficult task indeed, and one which would be expected to be performed better with increasing knowledge and experience.

CONCLUSIONS

Image reporting may, then, be argued to have three main elements – understanding what is in the images, understanding what is in the patient, and understanding the context in which the procedure was carried out. The first element requires knowledge of the science and technology of clinical imaging. The second element requires a grasp of anatomy, physiology and the mechanisms and manifestations of pathology. The third element requires an understanding of medical and surgical treatment and the objectives of clinical care.

Acknowledgement

I am grateful to Mrs J.M. Fletcher, Mr I. Crawshaw, Mr G. Culpan, Mrs L. Ford, Dr E. Robertson, and Mrs M. Wiggins for their contributions to the report analysis.

COMMENT   Richard Price, *Hatfield*

It was, of course, the vexed subject of reporting that dominated the relationship between radiographers and radiologists for, at least, the first 30 years of this century. Since then, the question of reporting, especially, plain films has continued to simmer below the surface but over the last five years the matter has been brought to the fore once again and not without some controversy.

The controversy of who should report is identified by Dr Robinson in his introduction. He also recognizes the parallel issues of whether reports are always necessary; whether reporting should form an integral part of an imaging procedure; or whether it should be a separate activity. All of these are important matters of substance and whilst he does not address these issues directly, the chapter does make an important contribution to the current debate on reporting. By considering the question – what is a report? – in a refreshingly neutral fashion, the chapter forms a basis for radiographers to consider their position in the current debate.

Radiographers have operated 'red dot' systems or variants, mainly in accident and emergency departments, for a number of years. Nuttall (1995) described one such system and showed its advantage in reducing the number of abnormalities missed by junior medical staff. Nuttall suggested that the system could be extended to include radiographers having a 'first read' to separate those films which need a radiological opinion from those that do not. Presumably, the first read is a report. McKay (1995) goes further and described a training programme for radiographers who will provide a radiographic report for casualty officers with the intention of extending this to general practitioner referrals. A pilot scheme, in Leeds and York, on reporting for radiographers has recently been concluded and a number of centres are offering courses for radiographers on reporting on the musculoskeletal system.

In the generic model proposed by Robinson for the structure of reports an assumption is made that a report is a 'series of data elements' that answer four questions:

1. What was done? – a description of the technique and its variations.
2. Did anything go wrong? – a description of the limitations of the procedure and any adverse events.
3. What did the procedure show? – an account of the findings.
4. What should be done next? – recommendations for subsequent imaging or therapeutic intervention.

A close analysis of these questions reveals nothing that is alien to the practice of radiography. Applied particularly to accident and emergency (A&E) radiography how many radiographers already provide such information, if only on an informal basis?

The first two questions: what was done?; did anything go wrong?; can only be answered directly by the person conducting the examination. In many cases this information is not available at first hand to the person making the report. The answers to the first two questions are relevant to 'what did the procedure show?' Radiographers already assess films on two levels, using technical and diagnostic criteria. Assuming a radiograph is technically sound, further decision-making requires the following steps:

Detection – is an abnormality present?
Localization – where is the abnormality?
Classification – what type of abnormality is it?

For radiographers the process may be largely informal, at present, but it underpins 'red dot' systems. What should be done next? could be recommended by the radiographer, for example further projections may be indicated or indeed a recommendation from a radiographer that a radiological opinion is required.

In discussing the model, Robinson goes on to argue that image reporting has three main elements which have particular requirements.

Main elements in reporting	Requirements
Understanding what is in the images	Knowledge of the science and technology of clinical imaging
Understanding what is in the patient	A grasp of anatomy, physiology and the mechanisms and manifestations of pathology
Understanding the context in which the procedure was carried out	An understanding of medical and surgical treatment and objectives of clinical care

The science and technology of imaging have always been core elements of radiographic education. Radiography is probably the one discipline to concentrate on this aspect. The application of science and technology to patients is the very basis of radiography. Anatomy, physiology and pathology are essential elements of radiography courses but they will probably require a specifically applied focus for courses in reporting. Finally, if practising radiographers, let alone those who will be involved in reporting in the future, do not have an understanding of the context in which a procedure was carried out then they should not be practising.

Robinson asks the crucial question; 'what is a good report?' Although an important indicator, a good report is more than just the speed of its production and return to the source of referral. A good report must be assessed on its findings, but Robinson indicates that audit of content is difficult in the absence of agreed standards against which performance can be measured. This is an area which has taxed those developing courses in reporting for radiographers because there is no 'gold' standard. Measures of accuracy do not differentiate between true positives and true negatives and measures of sensitivity and specificity based upon a consensus gold standard are not ideal. Robinson refers to the medicolegal position in the UK and states that the acceptable performance as 'that level reached by the majority of practitioners with similar experience and responsibilities.' But what is that level? A level presumably based around legal argument rather than objectively measured performance. The debate over radiographer's reporting may well expedite the setting of performance standards to the benefit of all those concerned, including the patient.

The College of Radiographers' Code of Professional Conduct (1994) and statement of infamous conduct from the Radiographers' Board of the Council for Professions Supplementary to Medicine (1995) place no impediment on radiographers making written comments on radiographs provided there is appropriate training. In ultrasound, especially in obstetrics, sonographers have had an important reporting role for a number of years and there is evidence to show that the role has been extended successfully into non-obstetric work (Bates et al. 1994).

The position of the Royal College of Radiologists (RCR) is made clear in its *Statement on Reporting in Departments of Clinical Radiology* (1995). It is that all examinations require a written report, presumably the RCR view of the ideal situation would be that a radiologist provides this report. However, the reality is that there is an insufficient number of radiologists and this is likely to continue. In its statement the RCR included the following:

> *After suitable training there may be no statutory impediment to a non-medically qualified person reporting a radiological examination and making clinical observations, but a person without a medical training cannot reasonably be expected to provide a medical interpretation.*

There has been animosity in the past over the division of labour and the establishment and consolidation of role boundaries. There is no evidence that radiographers wish to compete with radiologists and, in any case, do not have the medical background to do so. Radiographers, however, already make decisions on images and as Robinson concluded errors of perception can be reduced by training and errors of interpretation become less with experience. In the changing National Health Service, it makes sense to utilize and extend the skills of radiographers to include first line reporting, certainly of the musculoskeletal where many films go unreported. A reduction in abnormalities undetected by junior medical staff is an obvious benefit. Together, radiologists and radiographers can work to improve the service to patients.

References

American College of Radiology. Commission on Diagnostic Radiology (1992) *Index for Radiological Diagnoses*, revised 4th edn.

Anderson, E.D.C., Muir, B.B., Walsh, J.S., Kirkpatrick, A.E. (1994) The efficacy of double reading mammograms in breast screening. *Clinical Radiology* **49**: 248–251.

Bates, J.A., Conlon, R. M., Irving, J.C. (1994) An audit of the role of the sonographer in non-obstetric ultrasound. *Clinical Radiology* **49**: 617–620.

British Standard 5750 (BS 5750). London: HMSO.
College of Radiographers (1994) *Professional Code of Conduct*. London: The College of Radiographers.
Deyo, R.A., McNiesh, L.W., Cone, R.O. (1985) Observer variability in the interpretation of lumbar spine radiographs. *Arthritis and Rheumatism* **28**: 1066–1070.
Gjorup, T., Brahm, M., Fogh, J., Munck, O., Jensen, A.M. (1986) Interobserver variation in the detection of metastases on liver scans. *Gastroenterology* **90**: 166–172.
Gray, H.W., Pearson, D.W., Moran, F., Bessent, R.G. (1984) Reporting of ventilation perfusion images for pulmonary embolism: accuracy and precision. *European Journal of Nuclear Medicine* **9**: 151–153.
Kundel, H.L. (1989) Perception errors in chest radiography. *Seminars in Respiratory Medicine* **10**: 203–210.
Markus, J.B., Somers, S., O'Malley, B.P., Stevenson, G.W. (1990) Double-contrast barium enema studies: effect of multiple reading on perception error. *Radiology* **175**: 155–156.
McKay, L. (1995) Radiographic reporting in diagnostic imaging. In Paterson, A., Price, R. (eds) *Current Topics in Radiography*, No. 1, pp. 52–58. London: W.B. Saunders.
Nuttall, L. (1995) Changing practice in radiography. In Paterson, A., Price, R. (eds) *Current Topics in Radiography*, No. 1, pp. 28–38. London: W.B. Saunders.
Pauli, R., Hammond, S., Cooke, J., Ansell, J. (1996) Radiographers as film readers in screening mammography: an assessment of competence under test and screening conditions. *British Journal of Radiology* **69**: 10–14.
Robinson, P.J., Fletcher, J.M. (1994) Clinical coding in radiology. *Imaging* **6**: 133–142.
Royal College of Radiologists (1995) *Clinical Radiology Quality. Specifications for Purchasers*. London: Royal College of Radiologists.
Sheft, D.J., Jones, M.D., Brown, R.F., Ross, S.E. (1970) Screening of chest radiograms by advanced Roentgen technologists. *Radiology* **94**: 427.
The Radiographers Board (1995) *Statement of Conduct*. London: The Council for Professions Supplementary to Medicine.
The Royal College of Radiologists (1995) *Statement on Reporting in Departments of Clinical Radiology*. London: Board of Faculty of Clinical Radiology.
Yerushalmy, J., Harkness, J.T., Cope, J.H. et al. (1950) The role of dual reading in mass radiography. *American Review of Tuberculosis* **61**: 443–463.

8
Ethical Dilemmas in Radiographic Research

Val Challen

INTRODUCTION

Radiography aspires to professional status and to recognition by others. However, to achieve this it needs to ensure the provision of benefits to patients that it 'professes' to be able to provide (Gillon, 1994). One way of substantiating this is through the firm underpinning of practice through research, evidence-based practice being the sign of maturity of a profession.

PROFESSIONALISM, ETHICS AND RESEARCH

Research is seen by some to be an ethical requirement of a scientist (Wall, 1989) as it advances knowledge which ultimately benefits others. Healthcare activity is ethical in principle in that it applies science to human beings (Wright, 1987). Thus, radiographers have a dual ethical responsibility to be involved in research. With research activity ethical issues arise which need to be addressed and a balance has to be reached between the scientific requirements of methodology and human rights and values which may be threatened by the research (Kimmel, 1988). The code of professional conduct may provide some guidance by acting as a stimulus to moral thinking (Chadwick and

Todd, 1992) and by raising awareness of the ethical issues which need to be addressed (Henry, 1995). But what a code of conduct cannot do, however cogently expressed, any more than the application of a particular ethical theory can, is to provide the answers to moral dilemmas. Examining different ethical theories, however, can assist in the understanding of what data and what questions are relevant in any situation involving an ethical decision (Redmon, 1986).

The two major types of ethical theories that have dominated Western philosophy and on which research choices may be made are teleological and deontological theories. The difference between these two theories will depend on whether consequence (telos) or duty (deon) is considered to be the primary consideration in determining the moral worth of an action.

THE USE OF CONSEQUENCE TO DETERMINE MORAL WORTH – UTILITARIANISM

The best known consequence based theory is that of utilitarianism formulated in modern times by Jeremy Bentham (1784–1832) and John Stuart Mill (1806–1873). The principle of utility presented by Mill stated that an individual ought to do that act which promotes the greatest good, happiness or satisfaction to most people or in other words, produces the best possible set of consequences. In this formulation it is important to remember that the 'right' conduct is a matter of calculation (Bentham's felicific calculus) with each person's 'good' counting the same as every other person's 'good' (Kimmel, 1988; Wright, 1987).

If this aspect of utility is considered, then in a research situation it might be morally justifiable to deceive participants of the 'real' nature of the research in which they agree to participate as, in the long term, a larger number of people may benefit from the results of such activities. The scientific ends (further knowledge gained) may justify the means (deceit, manipulation or risk of harm) (Baumrind, 1985).

The major problem with this approach is that consequences can only be predictions and can never be certainties so the certainty of future benefits arising from a particular research programme, can never be assured (Wright, 1987). In addition, occasionally good may be gained at the expense of others thus leading to the problem of justice.

Health screening programmes, such as breast screening or routine antenatal ultrasound screening, take a utilitarian approach and sometimes fail to identify the psychological or other harm that can be done in screening a presumed healthy population. What utilitarians argue is that what is moral is dependent on the act and not on the motives of the agent; a 'good' end will reflect social utility and hence justify the means (Henry, 1995).

THE USE OF DUTY TO DETERMINE MORAL WORTH – KANTIANISM

Deontological theories are based on obligations often referred to as duties and the best known one is that of Immanuel Kant (1724–1804) who formulated the notion of the categorical imperative. This is a process for determining whether a proposed rule reflects a moral duty and defines principles of behaviour to be adopted in all situations as a matter of duty (Wright, 1987). Consequences are considered mostly, though not totally, irrelevant and achieving a 'good' end will never justify the means.

At first glance it would appear that a Kantian would have difficulty in involving people in research unless it was of a therapeutic nature relevant to the individual patient. Kant's approach is to suggest that people should be treated as ends in themselves and not *merely* as means to an end, as they have worth which is independent of a researcher's requirement of them. One interpretation of how an individual could become a research participant in Kantian terms is provided by Jonas (1969) who suggested that if the goals of the research are the same for both participant and researcher then the subject is not being *merely* used. How this can be established is somewhat problematic, particularly for the researcher.

GUIDING MORAL PRINCIPLES

There are problems with both the theories outlined, not least in connection with research activity. However, it is important to remember that an ethical theory is a constantly evolving framework (Wright, 1987), an ideal which can be used as a guide in the practical situation involving decisions which affect people. Again, a balance must be struck between the needs of the profession and the rights of patients which must be respected. Indeed one of the central tenets of professional practice should be respect for persons (Keyserlingk, 1993; Henry, 1995).

The principle of respect for persons involves the recognition that human beings have their own interests and opinions which need to be sought when decisions of a moral nature are being taken about them. Respect for persons is an extension of, and encompasses, the four prima facie moral commitments which are respect for autonomy, beneficence, non-maleficence and justice. These principles can be used to raise awareness where there may be a moral issue at stake and can guide in decision-making in a variety of contexts. In particular, they assist in identifying the moral dilemmas which may occur at the various stages of the research process.

RESEARCH STAGES AND MORAL ISSUES

Cartwright (1983) outlined six stages of a survey that, potentially, involved ethical considerations; see Table 8.1.

Table 8.1 Cartwright's six stages

1. Deciding to do the survey
2. Sampling
3. Data checking
4. Questioning contents
5. Data processing
6. Representation/publishing of results

Cartwright's stages can be used as a template to examine issues of an ethical nature which may arise in any research project – see Table 8.2.

Table 8.2 Stages of a research programme where ethical issues arise

1. Research area and appropriate methodology
2. Participant selection and assignment
3. Methods of data collection
4. Data processing
5. Representation, conclusion and publishing results

All research involving human subjects as patients, or in some cases as healthy non-patients, is considered by the Royal College of Physicians (RCP) to be medical research – whether it is carried out by nurses, doctors or paramedical professions (Royal College of Physicians, 1990) – and thus should be subject to ethical review by committee before commencement. The Declaration of Helsinki and its subsequent revisions was made to ensure that all medical research conforms to certain principles. These include the requirements that research should be humane in purpose, scientific in practice, and properly supervised. The competence of the researcher in being able to carry out the investigation and the ability of the supervisor are, therefore, ethical considerations related to respect for autonomy, beneficence and non-maleficence.

STAGE 1: RESEARCH AREA AND APPROPRIATE METHODOLOGY

The first moral dilemma to consider is the reasoning behind the research and the choice of research topic. If it is a scientific, professional and ethical duty to undertake research to improve practice and gain knowledge, then the very act of undertaking a research investigation involving human subjects for the sole purpose of a course requirement with no intention of widely disseminating the results could be construed as ethically dubious practice. Before commencing any research the questions that must be asked are 'whose interests are at stake in this piece of research?' Is it the individual researcher, the profession or the patient, actual or potential? Further questions that must be asked are 'would anyone feel threatened by the type of research, the questions asked or the potential findings?' 'Is there a likelihood that harm could occur, either physical or psychological, from the research?'

These are all questions of an ethical nature and involve all four moral commitments that radiographers have to people in their care. Autonomy, described by Gillon (1994) as deliberated self-rule, is the thread running throughout all the stages of the research process and should cause reflection on the moral issues that arise. Respect for autonomy requires consultation with patients (and others) and their agreement obtained before any procedure may be carried out on them. Voluntary informed consent should be seen as the central norm governing the relationship between researcher and the research participant (Kimmel, 1988). But, just as in undertaking an examination on a patient, problems may arise as fully informed consent would require discussion of very detailed and potentially unnecessary or unknown factors with participants. The process of informed consent is therefore often modulated by assessing the probability and degree of risk to the participants (Kimmel, 1988). However, if an ethnographic study is contemplated then to reveal the exact nature of the study to potential participants may negate the validity of the research leading also to problems of reflexivity. Researchers must take the decision for themselves as to whether covert participant observation can be seen as a violation of the principle of informed consent and thus autonomy. Subjects may have had no opportunity to give consent and may be kept in ignorance of, firstly, being watched and, secondly, the purpose of the research. Radiographers contemplating covert participant or non-participant observation should seek guidance from the ethical codes of their profession and the hospital ethics committee (Bulmer, 1982).

So choice of areas and of methodology consistent with the demands of the research question are important ethical considerations and should involve decisions being made by the researcher before commencement and may, in some cases, lead to the conclusion that this particular research programme should not proceed – a difficult decision for the individual to take.

STAGE 2: PARTICIPANT SELECTION AND ASSIGNMENT

In radiographic research the issue of selecting and assigning research participants may raise a number of ethical issues. Patients and/or relatives visiting the hospital/clinic have attended for a specific purpose which generally does not involve being part of a research study. However, on attendance for investigation or treatment they may be recruited onto a research programme; in this situation there is a 'captive audience' who are in danger of being coerced into a programme which the researcher may consider to be beneficial to 'patients' or the 'organization'. Where does respect for autonomy fit in here? The patient has the right to refuse but may be fearful that refusal will jeopardize his/her 'treatment' in the future. In addition, recruiting people who have a limited understanding of the language or true awareness of their role must be construed as unethical.

In the area of experimentation and randomized controlled trials the ethical issues are more pronounced if assignment of patients to a control group or experimental group is under consideration. The issue of randomization puts forward two opposite and distinct dilemmas. Firstly with regard to the 'untreated' control group – critics would argue that it is unethical to deny to people potentially beneficial treatment especially if the trial is a lengthy one (e.g. the tamoxifen trials carried out over a ten year period). Secondly for the 'treated' experimental group – critics would argue that it is unethical to subject participants to potentially harmful 'treatments'. This would be seen as using people as a means to an end (Kidder and Judd, 1986). One example in radiography would be assigning one group of patients to full information prior to a radiological examination with the control group being given no information. A compromise is often made to ensure that the control group receives some minimal information rather than no information. The prima facie issue here is one of justice.

Respect for persons requires respect for autonomy. Gillon (1994) asks who or what counts as an autonomous agent? Does it include the very young, severely mentally ill people and elderly people who may be mentally impaired? Some writers have argued that if children cannot give informed consent then they should not be made the subject of non-therapeutic research – as it negates the principle of respect for persons (Ramsay, 1970). A Kantian view would also rule out any research on children which does not have a direct bearing on the child's condition, i.e. therapeutic research. McCormick (1974), however, suggests that proxy consent by parents on behalf of children does not 'use' children, as the 'good' of children can be furthered through non-therapeutic research and that they 'ought to do so'.

Redmon (1986) disagrees with McCormick's view but puts forward a cogent argument that if it can be reasonably expected that the child on becoming an adult will agree with the goals of the research, and the possibility of harm is slight, then non-therapeutic research on children is permissible.

STAGE 3: METHODS OF DATA COLLECTION

Confidentiality and privacy are important ethical issues relating to respect for persons and for their autonomy and are related to this stage of a research programme as well as the remaining stages. Confidentiality is seen as one dimension of privacy with privacy referring to persons and confidentiality to data and information (Kimmel, 1988).

Clause 1.1 of the Code of Professional Conduct for Radiographers states 'radiographers must hold in confidence any information obtained through professional attendance on a patient' and professional attendance, of course, includes research. Westin (1968) defined privacy as 'the claim of individuals, groups or institutions to determine for themselves when, how and to what extent information about them is communicated to others'. Confidentiality may, therefore, include such information gleaned by a researcher which the research participant would rather not be known or recorded. One example may be information related to the subject's economic or mental health status and/or personal relationships. Researchers must be careful to guard confidential information from others, particularly those not involved in the research, and care must be taken over methods of storage of information, to limit access.

One other ethical dilemma related to the privacy of research participants is to avoid the use of painful or embarrassing questions which may cause anxiety and thus go against the commitments of nonmaleficence and beneficence. An example would be in an antenatal ultrasound survey probing patients about previous pregnancies and, possibly, terminations – spontaneous or otherwise.

Careful vetting of potentially problematic questions is an ethical activity as, too, is ensuring that the number of questions is limited to only those that are clearly necessary for the present study.

Anonymity must be assured such that no one must be able to recognize an individual by name or inference. For example, a department survey which as part of the data requests gender and status may enable individual male practitioners to be identified clearly in a female dominated department.

Covert participant observation can be seen as an invasion of personal privacy particularly if insinuation into a particular setting violates the rights of an individual to control his/her own sphere. Some commentators have argued, however, that covert methods of data collection do not harm the subjects of research if the exact identities of the people and of the research is kept concealed (Bulmer, 1982). Those seeking to use this type of data collection may be well advised to seek guidance. Debriefing participants may be seen as a necessary element in the research process but, in itself, it should not provide justification for unethical aspects of an investigation.

STAGE 4: DATA PROCESSING

Kimmel (1988) suggests that in addition to the collection of data ethical problems arise from the analysis of the data. One ethical dilemma that may arise is if the researchers' own theories are threatened by the data collected. Yet researchers have an obligation to observe and report all data completely and accurately. Babbage (1969) suggested that violations can occur and include 'cooking' the data (selecting data that fits into the hypothesis) 'trimming' the data (manipulating data to fit into the hypothesis) and forging the data (fabricating the data).

The use and manipulation of inappropriate or incompetently applied statistical tests to data collected may result in inaccurate or unreliable conclusions which may cause problems of reproducibility by other researchers; all these factors can be seen to be unethical in principle.

STAGE 5: REPRESENTATION, CONCLUSION AND RESULTS (PUBLISHING)

The researcher should ensure that he or she identifies the limitations of the research project and the results gained. This is an ethical requirement as it is not acceptable to suppress inconvenient or non-hypothesis supporting evidence.

With regard to the gaining of knowledge, another ethical issue arises as to whether the public is better served if the results are published as soon as possible in order to disseminate knowledge widely or whether it is preferable to wait until they are peer-reviewed before publication (Bermel, 1985). There should be no doubt that the practitioner should publish, this is an ethical requirement. The point is when should this occur.

If the research is funded through an agency the researcher should ensure that no pressure is applied (Kimmel, 1988) to present only that data which reflects the particular viewpoint of the organization responsible for funding. The researcher should document and clarify these issues with a funding agency before undertaking the research.

CONCLUSION

All stages of the research framework, including the design and methodology of the procedure, raise a myriad of ethical issues which need to be addressed by the researcher. She or he needs to be aware when there is a moral issue at stake. Research that is badly conceived, organized and executed is essentially unethical and is in danger of breaching the four basic prima facie moral commitments of respect for autonomy, beneficence, non-maleficence and justice.

There is no 'ethics of radiography' in the same vein as there is 'a nursing ethics' although the probity of the latter has been questioned by some writers (Melia, 1994). Is a set of ethical principles for radiographers conducting research with human participants required or can they be relied on to work within the RCP guidelines? A strong case could be made for either. In addition, student radiographers (undergraduate and postgraduate) have the ethical duty to seek the advice of their hospitals/trusts' ethics committees and to become fully conversant with the ethical principles and guidelines for research provided by their university. Supervisors of projects also need to ensure this happens.

COMMENT   Richard Price, *Hatfield*

Most practitioners at some time or another will be confronted with a difficult situation which will give rise to an ethical dilemma. The dilemma could relate directly to practice or to research, the latter being the focus of the chapter. Given the importance of ethical principles, ethics, is a subject that has not featured to any great extent within the radiographic literature. Val Challen, therefore, does a service to radiography in the UK by raising the subject in its own right.

The chapter recognizes the essential underpinning that ethics has for health-care activities. Ethics as defined in the *Oxford Reference Dictionary* (1992) is the science of morals in human conduct and as moral principles or code. Challen, cleverly uses ethical principles as the link between research, science, health care and professional status. For example, research is cited as being an ethical requirement of a scientist. Research is a tool of science and if properly conducted is a systematic investigation to establish new facts and conclusions. These should contribute to a body of knowledge which can be translated into practice. Healthcare, including diagnostic procedures, is said to be the application of science to human beings, and as such, there is an obligation to ensure that a course of treatment or investigation is based upon ethical principles. However, this does not mean that the establishment of what is ethical, or not, is necessarily an easy decision. Challen introduces ethical theory and models which can help in the understanding and analysis of fundamental problems that can lead to a resolution of dilemmas.

Therefore, the relationship between professional practice, status and ethical principles becomes clearer. Certainly, the theoretical link is straightforward; or is it? Challen makes the point that radiography aspires to professional status which it can achieve by research and hence evidence-based practice. There is an assumption here that this status has not yet been reached; this has been explored in a wider context in chapter 1 of this book. However, one cannot doubt the aspiration and indeed, it is supported by the aims of the Society and College of Radiographers which are to 'promote, study and research work in radiography . . .' The aims are supported further by the College of Radiographers in its policy document *A Strategy for Research* (1994) and the *Code of Professional Conduct* (1994).

Blane (1991), in identifying conditions that professions have in common, claims that 'professions espouse a code of ethics which prohibits the exploitation of clients and regulates intra-professional relations'. The College of Radiographers' Code of Conduct fits the description of an ethical code. The Radiographers Board (1995) meets its obligation as a statutory body by establishing a code of conduct that recognizes the principle of a 'self-imposed code of ethics' which defines the relationship between radiographer and patient and community at large which presumably includes research activities. The paragraph of importance here states that:

> *No registered radiographer should:*
> *by any act or omission do anything or cause anything to be done which he or she has reasonable grounds for believing is likely to endanger or affect adversely in a substantial way the health or safety of a patient or patients.*

Here the message is quite clear, but codes do not or cannot resolve ethical dilemmas. These must be a matter for those concerned, including patients and wider society if appropriate. But judgements and decisions must be made on an informed basis. Research is the vehicle for evidence-based practice but research must be rooted in valid ethical principles. This supports the inclusion of ethical theories and models within radiography curricula. Whilst there is evidence from an investigation undertaken at the University of Hertfordshire (Prime et al., 1995) that research is an important component of radiography courses the extent to which ethical theories and models are included is less certain.

The move to the higher education sector puts radiography research on a strong footing. Both undergraduate and postgraduate students have opportunities to conduct research which were not available a decade ago. There may not be any 'ethics of radiography' but as Challen has illustrated there are well-established theories to draw upon. It could be useful to call upon philosophical medical ethics that Gillon (1992) describes as the analytical activity in which the concepts, assumptions, beliefs, attitudes, emotions, reasons and arguments underlying medicomoral decision-making are examined critically.

Is the lack of any formalized radiography ethics because none have been proposed? Whether this will change in the future, only time will tell, but the chapter provides a sharp focus to begin the debate. It is up to the profession to demonstrate in a transparent manner its ability to translate research into evidence-based practice. There can be no disagreement with Challen's position that radiographers have a dual ethical responsibility to be involved in research.

References

Babbage, C. (1969) As cited in Kimmel, A.J. (1988) *Ethics and Values in Applied Social Research*. Newby Park: Sage.

Baumrind, D. (1985) Research using intentional deception: ethical issues revisited. *American Psychologist* **40:** 165–174.

Bermel, J. (1985) Two research studies, two views. *Hastings Center Report* **15:** 3–4.

Blane, D. (1991) Health professions. In Scambler, G. (ed.) *Sociology as Applied to Medicine*. London: Baillière Tindall.

Bulmer, M. (1982) *Social Research Ethics*. London: Macmillan Press.

Cartwright, A. (1983) *Health Surveys in Practice and Potential*. Oxford: OUP.

Chadwick, R., Todd, W. (1992) *Ethics and Nursing Practice*. Basingstoke: Macmillan.

College of Radiographers (1994) *A Strategy for Research*. London: College of Radiographers.

College of Radiographers (1994) *Code of Professional Conduct*. London: College of Radiographers.

Gillon, R. (1992) *Philosophical Medical Ethics*. Chichester: John Wiley.

Gillon, R. (1994) Medical ethics: four principles plus attention to scope. *British Medical Journal* **309:** 184–188.

Henry, C. (1995) *Professional Ethics and Organisational Change in Education and Health*. London: Edward Arnold.

Jonas, H. (1969) Philosophical reflections on experimenting with human subjects. *Daedalus*, Spring, 219–247.

Keyserlingk, E.W. (1993) Ethics, codes and guidelines for health care and research: can respect for autonomy be a multi-cultural principle? In Winkler, E.R., Coombs, J.R. (eds) *Applied Ethics; A Reader*, pp. 340–415. Oxford: Blackwell Scientific.

Kidder, L.H., Judd, C.M. (1986) As cited in Kimmel, A.J. (1988) *Ethics and Values in Applied Social Research*. Newby Park: Sage.

Kimmel, A.J. (1988) *Ethics and Values in Applied Social Research*. Newby Park: Sage.

McCormick, R.A. (1974) Proxy consent in the experimental situation. *Perspectives in Biology and Medicine* **18:** 1. Reprinted in Mappes, T.A., Zembaty, J.S. (1981) *Biomedical Ethics*. New York: McGraw-Hill.

Melia, K.M. (1994) The task of nursing ethics. *Journal of Medical Ethics* **20**: 7–11.

Prime, N.J., Higgs, A., High, J. (1995) *Curriculum Development in Radiography – Results of the Desk Research*. University of Hertfordshire (unpublished).

Radiographers Board (1995) *Infamous Conduct*. London: The Council for Professions Supplementary to Medicine.

Ramsay, P. (1970) *The Patient as Person*. New Haven: Yale UP.

Redmon, R.B. (1986) How children can be respected as 'ends' yet still be used as subjects in non-therapeutic research. *Journal of Medicl Ethics* **12**: 77–82.

Royal College of Physicians (1990) *Guidelines on the Practice of Ethical Committees in Medical Research Involving Human Subjects*. London: RCP.

Wall, A. (1989) *Ethics and the Health Services Manager*. London: King's Fund Publications Office.

Westin, A. (1968) As cited in Kimmel, A.J. (1988) *Ethics and Values in Applied Social Research*. Newby Park: Sage.

Wright, R.A. (1987) *Human Values in Health Care: The Practice of Ethics*. New York: McGraw-Hill.

9
The Patient's Charter and the Delivery of Quality Services

Marilyn Hammick

INTRODUCTION

Consumer choice has been, and continues to be, a leading tenet of modern-day Conservatism. It is inextricably linked with the market philosophy which now pervades public services in the United Kingdom. Conservative consumerism is epitomized by 'charterism', with the Citizen's Charter being followed by charters associated with particular state funded services (Cohen, 1994). The Patient's Charter (Department of Health, 1991) documented ten rights of the National Health Service (NHS) patient and nine standards of health care. (Hereafter the Patient's Charter and the rights and standards it addresses are referred to as the Charter, the Rights and the Standards.)

The major focus of the Charter is on the information about, and the opportunities for, health care that should be available to citizens. This fits into the narrow pattern of Conservative consumerism which excludes legal rights and rights of participation in decision-making (Harrison et al., 1992). The obvious power differential, between legal and decision-making rights and information and opportunity rights, demonstrates how little control is given to the patient by the Charter. The Patients' Association thinks the term *rights* unrealistic, preferring the phrase 'the right to expect' (Cohen, 1994).

Issues of quality have a history of being linked to the patient as a valued consumer (Moore, 1988). As the idea for the purchaser–provider split germinated in politicians' minds, the forecast of the 'rise of the consumer' as an 'increasingly important feature of the next 40 years', and of the need for health care to take account of this, was made (Moore, 1988). The Charter continues the tradition of many of the recent NHS changes by seeking to empower the patient as consumer. But many factors influence the ability of the patient to act as a consumer of the NHS. Patients can, and should, make choices about many aspects of their care but remain dependent upon the expertise of practitioners for the selection of some critical interventions on their behalf. Within radiography services ultrasonographers intervene on the patient's behalf when they make referrals for genetic counselling. In the contracting process, as Alaszewski (1995) highlights, general practitioners 'act as "surrogate" consumers . . . on behalf of their patients'. With financial considerations the driving force behind contracts, the perception that patients can make 'informed choices about their local hospitals' is 'little more than a myth' (MacAlister, 1994).

The danger in setting Rights and Standards is that patients may think, for example, that they are in the best place when they are seen within thirty minutes in outpatients but the reality can be very different. The experience of Moser (1994) who complains of the 'rude, brusque and impatient' manner in which she was treated during an outpatient consultation is unlikely to be unique. Moser (1994) claims that the 'true quality of care is deteriorating and patients are the victims'. Far from empowering patients the Charter has the potential to reduce them to items in the production line of consultation.

The power of patients to implement their Rights depends on an awareness of these, and a belief in the ability to exercise them. Evidence suggests that distributing the Charter to every household in the country has not been synonymous with widespread public awareness of it. Two years after publication, only 19% of a sample interviewed said they had read the Charter, only 24% recalled having seen a copy and 35% did not remember hearing about it (Cohen, 1994). The Royal College of Nursing (RCN) (1994) highlight the lack of knowledge about the Rights by 'most people' and Richardson et al. (1994) report that 84.4% of a group of people from ethnic minorities (n = 77) either had not heard of, or did not reply to, questions about, the Charter.

Tailor and Mayberry (1995) conclude that 'a credibility gap exists amongst patients', with less than half the subjects in a survey of hospital outpatients believing 'that (the Charter) will influence the standard of care they receive'. Many patients consider that money has been misspent on the paper associated with the Charter and would prefer to see its costs focused on direct patient care (W. Vlok, personal communication, 1995). This scepticism is shared by some healthcare professionals who feel that 'it does little to ensure consumers of health services get a better deal' (Reid, 1994).

The Charter does seek to achieve what many radiographers would want on behalf of their patients. But there are constraints alongside the praiseworthy conditions it offers. One of these relates to the influence of the medical profession on measures designed to implement the Charter. Another, operationalized by financial managers but ultimately in the hands of central government, is that implementation of the Charter is dependent upon 'circumstances and resources' (Department of Health, 1991).

These constraints raise several queries. There is the uncertainty about whether failure to implement the Charter will be explained away using the escape clauses built into it. The influence of doctors and financial managers, and the ability of other healthcare professionals to challenge this in order to implement the Charter, is important. Political questions about a possible hidden agenda behind the Charter are also relevant.

THE POLICY AGENDA

Enquiry about the political nature of the Charter is informed by testing it against a theoretical model of policy and highlighting issues about the government's aims for NHS charterism. Public policy is essentially a set of decisions, taken by political actors, and is about means and ends (Hill and Bramley, 1986). It is contingent, essentially in the case of healthcare policy, upon the prevailing economic climate. Hill and Bramley (1986) counsel that policy should not be treated as 'self-evident', nor as a constant but as 'continuing interplay through political action'. The Charter is essentially a set of government decisions about the public face of the NHS and will continue to be shaped by political and micro-political climates.

The political decisions about the Charter included the means of determining whether the Rights and Standards have been met. Where possible, the choice is of indicators that 'capture only one dimension', for example, standard times for being seen in outpatient clinics and waiting for an ambulance (Hogwood, 1992). This use of measures that are 'readily available and quantifiable' has the potential to conceal a lack of provision of quality patient care (Hogwood, 1992).

The statistical indicators of quantity provide the government with the means to its end. The public now have the opportunity to measure the service they receive against numerical national standards – a reductionist approach in which the evaluation of quality is largely neglected. However, this approach does not necessarily satisfy patients. Britten and Shaw (1994) report that patients attending accident and emergency departments identified a number of issues of importance to them which are omitted from the Charter. They conclude that the Rights and Standards are defined 'too narrowly' and that the 'Charter can only be meaningful . . . if it attends to those issues that patients consider important' (Britten and Shaw, 1994).

THE PUBLICITY AGENDA

The Charter, undoubtedly, has features of public policy and it is now an inherent part of the NHS. But is it just policy, or does it conceal political ambitions over and above ensuring that individual patients receive the very best in health care? Is the Charter also political publicity?

The Charter provides patients with criteria to evaluate aspects of the NHS, and the means for their opinions to be heard. One way of viewing this is as encouragement for the public to do what ministers have been doing for years (and continue to do); that is to call for greater efficiency and value for money. The responsibility for highlighting the failures of the health service, albeit at the level of the individual rather than society, is now with the public. The Charter provides a means of deflecting the origin of adverse comment away from the government who can then (benignly) intervene on the public's behalf.

Consumers of health services now have increased 'political importance' (Saks, 1995). The manner in which the Charter was written reveals the potential for a hidden agenda. Lamont (in Cohen, 1994) speaks of how it was written with little consultation with health groups, and the haste, for 'political reasons', with which it was sent out. There is strength in the view that poor reception of the Charter is due not only to the failure to sell it to the public but also to its political context (McSweeney, 1994). This context is heavily influenced by Conservative consumerism which induces scepticism by citizens about the effect of charters on their lives.

The Charter has also been the topic of wider polemic with Simpson (1993) writing of waiting lists as '(a) political battleground' linked to the massaging of what *being on a waiting list* now means. There is also a local micro-political climate in which Charter initiatives are set.

Cohen (1994) reflects that the Charter is 'politically driven', with 'performance-related pay depend(ent) upon meeting Charter targets'. It is at this level that health care professionals become linked into the politics of the Charter.

THE PROFESSIONAL AGENDA

Professional idealogies have influenced the agenda for health care in this country since the inception of the NHS. The power struggle between the professions, most dominantly the medical profession, and the politicians remains a significant part of that agenda. The restructuring of funding arrangements continues to play a major part in the state's pressure for control in this power struggle, and is now joined by mechanisms to enhance the power of patients (Alaszewski, 1995). Any success by the state to diminish the medical profession's power will alter the inter-professional balance of power.

Alaszewski (1995) points out that the traditional role of nurses will be transformed as the result of many of the Working for Patients changes to the NHS, promoting a complementary and independent role. The Charter highlights nurses as having specific responsibilities for patients emphasizing that nursing care is as much part of the health service as medical care. Radiography services will be affected by such changes and, at the same time, radiographers can be influential.

The named-nurse system can be difficult to implement within radiography services where many patients attend only as outpatients. This presents opportunities for radiographers to implement a named-radiographer system linking patients to a qualified radiography practitioner. Some patients requiring diagnostic procedures will pass quickly through the department and a named-radiographer may not have a major impact upon their care. However, for many others their relationship with the radiographer is a fundamental part of their experience.

Radiography practitioners have an important informational and support role before, during and after radiographic procedures. Extending the naming system to ensure that individual practitioners are linked to patients requiring similar examinations and treatments will develop the specialist role of each practitioner. Traditional ways of working may have to change if the Rights and Standards are to be met. Greater involvement of radiographers in procedures presently done by clinicians presents opportunities for increased efficiency.

With increasingly sophisticated technology the radiographer's role includes ensuring optimal use of all available resources. The aim is for high quality procedures, and the assurance that patients, their family and friends are supported from reception into the department, throughout all treatment or imaging procedures and in the follow-up period. Implementation of a system where a named-radiographer has primary responsibility for a particular patient, and has developed advanced practice knowledge and skills in respect of the radiographic procedure used, will enhance the quality of patient care.

A named-radiographer system should be part of future radiography services, with the Charter providing reason and purpose behind this radical change to professional practice. Transforming the allocation of work within radiography services from an equipment-related approach to one which is orientated towards patients has the potential to give radiographers increased independence as practitioners. Shifts in inter-professional power, likely to result from implementation of the Charter and other changes to the NHS, are part of an environment that will support role extension and the development of advanced practice fundamentals in the provision of a quality service.

RADIOGRAPHY SERVICES: THE DELIVERY OF QUALITY

Although there is no mention of quality in the rationale for the Charter, phrases used in the foreword such as 'High standards of the best', and 'services that provide clear and measurable benefits', indicate that the aim is for a quality service (Department of Health, 1991). The quality of radiography services is dependent upon many factors and can be achieved in many ways but the Charter has potential as a framework for achieving quality in aspects of radiography services. Other examples of similar management tools are quality circles and systems established to obtain British Standard 5750 certification.

Whatever the mechanism, working to provide a quality service is likely to mean that many of the Rights and Standards will be met. The question remains, however, whether meeting these means the provision of a quality service in all respects. The discourse of quality service provision is wide. The practice of quantitatively measuring achievement of Charter targets can surely only be a minor part of any claim to knowing about this achievement. Other, more important, aspects of the discourse include defining what 'degree of excellence' the service aims to achieve, and relating this to customer satisfaction (Luthert, 1992).

One implication of the Charter for those responsible for service provision is about underpinning the Rights and Standards that are quantified, with quality. Reality may mean, however, that the achievement of this is limited by the level of control available to change practice.

Radiography services, as an inherent part of health care provision, encompass a wide range of procedures. The role played by radiography practitioners is both unique and inclusive and this is an important facet of both the delivery and quality of radiography services.

Radiographers are responsible for aspects of patient management and care that relate specifically to radiographic practice. These procedures are not within the remit of any other healthcare practitioner. This is one basis of radiographers' professionalism. The radiographer is also a vital part of the healthcare team and participates in procedures that depend for success on the integrated functioning of that team. In different situations the members of a team, and its hierarchy, will vary.

Radiographers, therefore, have two roles in the delivery of a quality service. In one they have a large degree of control over the service being delivered. In the other control is shared. The influence of any one member can be attenuated by power differences within the team, and by traditional ways of working. As a result service quality may be less than optimal.

The Charter includes Standards which uphold the ethical basis of health care, to ensure respect for patients and protection of their autonomy. Many ways of showing respect and ensuring autonomy are within the control of individual radiographers. The Charter presents opportunities for radiographers to confirm that their working practices ensure that 'dignity and religious and cultural beliefs are respected' (Department of Health, 1991).

The introduction of new x-ray gowns that fit all patients, even after laundering, at the Royal Preston and Sharoe Green Hospital is a local initiative by the X-Ray Quality Circle that implements Standard 1 (NHSME, 1993). The manner in which requests from patients to be treated by staff of the same gender are treated will determine whether Standard 2 is met. This, then, becomes one of the factors that need to be considered by radiotherapy service managers when allocating staff to a particular treatment unit team. One professional responsibility of radiographers is the provision of information about radiographic procedures in an appropriate manner. The development of radiology (sic) patient information leaflets and audio tapes by radiography staff at Harrogate Health Care NHS Trust illustrates a way of meeting Right No.2 (NHSME, 1993).

Unlike the examples given above other Rights and Standards are beyond the control of individual practitioners. In addition to the power of colleagues, many aspects of a quality service will depend upon resource allocation. For example, the adaptation of mobile breast screening units to enable access by wheelchair users to meet Standard 2 is likely to be influenced by the funding available within provider units. At this level radiography managers have a role in the distribution of funds for Charter initiatives. Provider funds are contingent upon the cost of contracts placed by purchasers, and overall Exchequer funding of the NHS will have the final say. Conviction that the Charter is public policy carries with it the expectation of support by sufficient state funding. If it is also, perhaps only, political publicity then it would not be surprising to find that the rhetoric fails to include a financial commitment.

CONCLUSION

The Patient's Charter is an important part of the NHS market. It will continue to play a role in controlling the delivery of all health services, including radiography. The willingness of individual radiographers, and the profession, to take forward the opportunities to extend their role and responsibilities offered by the Charter will influence the quality of services they provide. Working within, not only the narrow focus of the Charter, but also the broader discourse of quality service provision, radiographers can play a significant part in developing patient-centred services that are meaningful to all.

References

Alaszewski, A. (1995) Restructuring health and welfare professions in the United Kingdom: the impact of internal markets on the medical, nursing and social work professions. In Johnson, T., Larkin, G., Saks, M. (eds) *Health Professions and the State in Europe*, pp. 53–74. London: Routledge.
Britten, N., Shaw, A. (1994) Patients' experiences of emergency admission: how relevant is the British government's Patient's Charter. *Journal of Advanced Nursing* **19**: 1212–1220.
Cohen, P. (1994) Passing the buck? *Nursing Times* **90(13):** 28–30.
Department of Health (1991) *The Patient's Charter*. London: HMSO.
Harrison, S., Hunter, D.J., Marnoch, G., Pollitt, C. (1992) *Just Managing: Power and Culture in the National Health Service*, p. 136. Basingstoke: Macmillan.
Hill, M., Bramley, G. (1986) *Analysing Social Policy*, pp. 137–140. Blackwell: Oxford.

Hogwood. B.W. (1992) *Trends in British Public Policy*, pp. 4–5. Buckingham: Open University Press.

Luthert, J. (1992) Quality assurance and standards of care, *European Journal of Cancer Care* **1(2)**: 34–37.

MacAlister, L. (1994) NHS league tables: does a 5-star rating indicate 5-star care? *British Journal of Nursing* **3(13)**: 647–648.

McSweeney, P. (1994) Healthy remedy or sick joke? *Nursing Standard* **8(42)**: 20–21.

Moore, J. (1988) A Health Service for people. In *People as Patients and Patients as People*. London: Office of Health Economics.

Moser, K. (1994) Letter to the editor. *The Guardian*, 30 November.

NHS Management Executive (1993) *The A–Z of Quality A Guide to Quality Initiatives in the NHS*. London: Department of Health.

Reid, T. (1994) Editor's note. *Nursing Times* **90(13)**: 28.

Richardson, J., Leisten, R., Calviou, A. (1994) Lost for words. *Nursing Times* **90(13)**: 31–33.

RCN (1994) *Unchartered Territory: Public Awareness of the Patient's Charter*. London: RCN.

Saks, M. (1995) The changing response of the medical profession to alternative medicine in Britain A case of altruism or self-interest. In Johnson, T., Larkin, G., Saks, M. (eds) *Health Professions and the State in Europe*, pp. 101–115. London: Routledge.

Simpson, I. (1993) How to avoid a waiting game; Health Insurance; Your Health, Scotland. *Sunday Times*, 13 June.

Tailor, H., Mayberry, J.F. (1995) The Patient's Charter: A survey of hospital out-patients views of their rights and ability to exercise them. *Social Science in Medicine* **40(10)**: 1433–1434.

10
Patients' Perceptions of Radiotherapy Treatment

Christine Mackenzie

INTRODUCTION

It is estimated that one in three people will develop cancer during their lifetime. The incidence increases with age, with 70% of all new cancers being diagnosed in the over 60s (Cancer Stats., 1993). Cancer is responsible for 24% of all deaths (CRC, 1992). It is therefore understandable that cancer evokes dread and fear. Approximately a quarter of women take more than three months to consult their doctor and a reason for this delay is fear of diagnosis (Green, 1976).

A number of studies have looked at the prevalence of anxiety in cancer patients. One of the most frequently cited is that of Derogatis et al. (1983), which found that 101 cancer patients from a population of 215 could be labelled psychiatric 'cases'. This is a prevalence rate of 47% in a cancer patient population and is three times higher than the general population. The nature of the various treatments given to cancer patients further increases their anxiety. Approximately 50–60% of all cancer patients receive radiotherapy (Crosson, 1984). In their study on mastectomy patients, Fallowfield et al. (1986) asked 'Looking back over this past year, can you pick out one period that was worse than any other?' Most women stated 'between finding the lump and hearing the diagnosis, closely followed by their experiences during radiotherapy'. There is increasing evidence that patient satisfaction is a factor of quality of care evaluation (Fitzpatrick, 1991) and this chapter will look at how patients perceive their radiotherapy treatment.

TREATMENT

There is no doubt that fears and misconceptions about the nature and aims of radiotherapy cause patients distress. In a study by Peck and Bolland (1977), two-thirds of their fifty patients remained anxious throughout their treatment. The most feared side-effects were burns (72%), scars (54%) and pain (54%).

Gyllenskold's (1982) study of breast cancer patients cites 'being radioactive', 'worries about damage to healthy tissue' and 'permanent genetic damage' as common misconceptions which increase patients' distress.

In recent years there has been much publicity about the effect of ultraviolet radiation on the skin and how it contributes to skin cancer. Similarly, the catastrophe at Chernobyl gave the world a considerable scare. For patients, this causes an anomaly. How can something which causes cancer, cure cancer? Some patients also feel that radiotherapy is only given to palliate symptoms and their surgery has therefore failed.

In addition to all these factors, other considerations can increase a patient's distress. Patients frequently have to wait for a considerable time in rather drab buildings, often in the company of the obviously sick and dying. This makes it increasingly difficult for them to sustain denial – a major coping strategy. News in the media about incorrect dosages has not helped patients' fears, particularly as radiation can neither be felt or smelt. For some, the whole experience is extremely threatening and anxiety levels are high among patients undergoing radiotherapy. In a study carried out by Young et al. (1992), 44% of radiotherapy patients had high anxiety levels at simulation.

Mitchell and Glickman (1977) studied patients having radiotherapy treatment and found that 80% would not discuss their emotional problems with their referring doctor or radiotherapist. Is this because patients feel that doctors and especially radiotherapists are highly trained 'technical experts' and they do not want to 'waste' their valuable time on their emotional problems? Cassileth et al. (1980) stated that radiotherapy patients felt poorly informed and desired more information, especially from their radiotherapist.

In a questionnaire, Anne Eardley (1986) asked patients what aspect of radiotherapy came as a surprise to them. The most popular response was that it was painless (43%). 'I was wondering what it would be like – whether it would feel hot, with the X-ray going through – whether there'd be any burning' (Eardley, 1986, p. 54).

Patients (29%) were also surprised at how quick the actual treatment was – 'it was all over in a matter of minutes – I was surprised it was so quick' (Eardley, 1986, p. 54). Another aspect referred to by patients was the fact they were left totally alone during treatment. 'They put the lights on, and out they went – they ran for the door. You didn't know what the hell was going to happen' (Eardley, 1986, p. 54).

These examples illustrated that patients were not receiving enough correctly focused information before their treatment has started. Does this continue throughout their treatment?

SIDE-EFFECTS

Are patients sufficiently prepared for side-effects? The literature tells us that the side-effects of radiotherapy further increase a patient's anxiety (Andersen et al., 1984). Anne Eardley (1986) found that 54% of her head and neck patients were not properly prepared. One quarter of her population thought that their side-effects would get worse after they left hospital, 40% were uncertain and 24% did not expect them to continue. At the end of treatment, Mackenzie et al. (1996) found that 59% were suffering from side-effects as a result of their radiotherapy treatment and, four weeks later, this had dropped to 47%. Some patients specifically mentioned 'I would have liked more information on the after-effects' and 'when completed, you could be told what to expect until you see the Consultant some weeks later'. A large percentage (43%) would have liked someone from the radiotherapy department to have telephoned them between the end of treatment and their appointment with the consultant.

Lack of communication could be causing patients to focus too closely on their side effects. In a study carried out by Parson et al. (1961) (which today would not be given ethical approval), patients were given 'sham' radiotherapy treatment. It was found that 75% experienced nausea and fatigue. This highlights the complex interaction between the physical and the psychological. Well informed patients are better able to cope both physically and emotionally. Leventhal (1986), Nerenz (1986), Leventhal et al. (1986) and Nerenz et al. (1986) have suggested that some side-effects can cause as much distress as the cancer itself.

INTERVENTIONS

Do interventions help patients to cope with psychological distress? Holland et al. (1979) carried out a study in which women patients were randomly assigned to either an intervention or a non-intervention group. The intervention group had a tour of the department which included talks on the procedures involved by the staff who would treat them, followed by a question and answer session. These women were found to be less anxious on their first visit. A less disruptive method of giving patients information could be by video presentations. Rainey et al. (1985) divided patients into low and high information groups. The low information group received only a booklet, whereas the high information group were shown a video with information about the staff, treatment, common misconceptions etc. Staff were present to answer any problems. Those in the high information group had less anxiety and mood disturbance when they arrived at the radiotherapy department and this continued to the end of their treatment.

Anxiety and depression have been reduced by psychotherapy (Spiegel, 1994). It has also been shown to be cost-effective. Mumford et al. (1984), using meta-analysis, showed a fall in medical services and a 1.5 day reduction in occupancy, when patients were offered psychotherapy. On average, extremely depressed and anxious patients had both a longer stay and higher costs than those with low levels of psychological distress (Levenson et al., 1990). Other interventions have shown reduction in symptoms such as pain (Spiegel and Bloom, 1983) and nausea (Morrow and Morrel, 1982). Meyer and Mark (1995) undertook a meta-analysis in psychosocial interventions with cancer patients and found that 'they have a positive effect on emotional adjustment, functional adjustment and treatment – and disease-related symptoms in adult cancer patients' (Meyer and Mark, 1995, p. 104).

A number of cheaper alternatives to psychotherapy exist. Patients (45%) prior to their treatment felt they would like to speak with a patient who had already undergone radiotherapy treatment (Mackenzie et al., 1996). Patients also felt the need for local support groups. Though research needs to be carried out on their effectiveness.

Many palliative radiotherapy patients are, however, old, frail and very ill. They have to spend all day waiting either for transport and/or radiotherapy for treatment that lasts only a few minutes. How effective is radiotherapy in controlling pain, when compared with pain relief with drugs? Maybe policies on single fraction treatment should be reconsidered so that patients, as recommended under the Calman Report, are ensured 'the best quality of life'. In a Norwegian study by Kaasa et al. (1993), 277 palliative radiotherapy patients were tested before their radiotherapy treatment. They concluded that the most distressed patients were those in the most pain with poor performance status. A total of 69% of the 247 patients who agreed to participate in the study reported a high level of psychological distress. This is approximately five to eight times higher than a normal population. Are these palliative patients routinely screened and, if necessary, offered extra psychosocial support? Jane Graydon (1988) with a sample of 79 patients found that those patients who were anxious and tense at simulation tended to have poor functioning following their treatment.

Calman further recommended that 'psycho-social aspects of cancer care should be considered at all stages'. Therefore, with 40% of patients exhibiting psychological distress at simulation (Maher et al., 1996), should not all patients arriving for simulation in a radiotherapy department be routinely screened, as part of their routine care, with a reliable validated questionnaire such as the HADS (Zigmond and Snaith, 1983) using a high cut-off point to allow for radiotherapy patients natural stress?

As patients become more informed about treatments such as radiotherapy, doctors may experience increasing pressure to justify their actions. Health professionals need to be sensitive to patients' needs and allow patient participation in treatment decisions. Is this discussed with patients in an honest fashion? Perhaps, the strategy should be changed and patients actively involved in decision-making. Patients who made their own informed decision as to whether they should have lumpectomy, mastectomy and/or radiotherapy experienced less immediate and long-term distress than those who did not (Morris and Royle, 1988).

CONCLUSION

It is now time for health professionals dealing with radiotherapy patients to reassess patient management. Communication and information have been shown to be effective in reducing psychological distress in radiotherapy patients. Are radiotherapy departments, however, finding the resources required, as these involve not just time but also education and training. Surely, this should be a top priority. There is much talk but little action. There are some departments who do cater for the physical and psychological care of the patient, but all too few. Many hospitals are now giving literature to patients when they first arrive or prior to simulation. How many of these hospitals have evaluated such material to demonstrate that it is effective? Does it answer patients' queries? How many patients are being given information on what to expect once their treatment has ended?

For many years the medical profession hid the word 'cancer' from patients, fearing that patients would not be able to 'cope'. However, the evidence has proved the contrary. Patients are better able to cope with these situations if they are well informed and are told in a sensitive manner (Cohen and Lazarus, 1979). It is rather that the health professionals have not been given the necessary training in communicating bad news (Maguire, 1990). Vachon and Conway (1989) reported that, though physical effects altered survivors' well-being, the psychological and social effects were of greater importance and attention to these seemed to be lacking.

In conclusion, many radiotherapy patients find their treatment frightening, distressing and time-consuming. The management of radiotherapy patients should be multidisciplinary. With the right focused information and support, cost-efficient and cost-saving treatment can be given to patients. Hippocrates said 'healing is a matter of time, but it is sometimes also a matter of opportunity'. It is to be hoped that, by the 21st century, patients will be given the maximum opportunity.

COMMENT   Audrey M. Paterson, *Canterbury*

As Mackenzie demonstrates, she is not the first to advocate the benefits of appropriate psychosocial care for people suffering from cancer, and it is heartening that the Chief Medical Officer (CMO) has accorded such care some importance. There is also evidence that at least some departments of oncology are developing comprehensive services to support patients on their 'cancer journeys' (Young, 1995), although the nature and extent of these varies considerably, and especially in relation to the roles and responsibilities of therapeutic radiographers (Paterson, 1995). The changes to cancer care heralded by the Expert Advisory Group on Cancer to the Chief Medical Officers of England and Wales (EAGC), perhaps, provide a timely opportunity for radiographers to review radically their fundamental roles in cancer care. It could be argued that for too long their practice has been rooted in a highly technological, biomedical model of care to the detriment of the human aspects of care.

The proposed new structure for cancer services proposes three levels of care with primary care as the focus, supported by cancer units and cancer centres. It is suggested that radiotherapy will take place only (or mainly) within the cancer centres although it is recognized that some cancer units may also have radiotherapy facilities. For therapeutic radiographers this poses a potential dichotomy. They may entrench themselves yet further in their traditional, medically-oriented, medically-dominated model of practice envisaged by the EAGC. Alternatively, they may seize the moment to redefine, defend and justify a much broader cancer treatment and care role – a role that places their vital skills in the optimization and delivery of high quality radiotherapy within the continuum of care that commences at, or prior to, diagnosis and continues to beyond complete cure or death.

Some will argue that such a role is not possible, and others will suggest that it is not desirable. However, in doing so, they deny the evidence. Therapeutic radiographers are the only health care professionals whose entire training is devoted to the treatment and care of people with cancer. Over half of all cancer patients will undergo radiotherapy as part of their treatment and management and, hence, will interact with radiographers for significant periods of time and at critical points in their disease. Indeed for those cancers treated radically, the very nature of the treatment regimes demand that it is radiographers with whom patients will interact principally for the duration of radiotherapy. Is it not, therefore, appropriate for therapeutic radiographers to place radiotherapy in the continuum of care offered to patients with cancer; and to assert their roles as key providers of that continuum of care?

Of course, turning rhetoric into reality is not easy, but it is possible for radiographers, especially those involved in managing services, to consider what is feasible, achievable and realistic within their own centres. The EAGC suggests that 'a crucial issue' is the development of appropriate contracting for purchasing cancer services. It also recognizes that this is neither an easy or a quick process. Nevertheless, it does provide significant opportunities for radiographers to demonstrate their pivotal role in cost-effective, holistic patient management that spans the proposed three levels of cancer care.

The practice of all health-care professionals, including radiographers, is under much scrutiny at present, and professions must increasingly demonstrate their value to the public they serve. Therapeutic radiographers have the opportunity to shape the future of cancer care and, thus, the future of their profession. They can do so positively, within the context of the inter-professional cancer care team and, in doing so, they can realize the hope of Mackenzie – that by the 21st century cancer patients will be given the maximum opportunity to overcome their disease.

References

Andersen, B.L., Karlsson, J.A., Andersen, B., Tewfik, H.H. (1984) Anxiety and cancer treatment: Response to stressful radiotherapy. *Health Psychology* **3:** 535–551.

Calman, K. (1994) A policy framework for commissioning cancer services. Consultative Document.

CRC (1992) *Cancer Research Campaign's Fact Sheet 5.*

Cancer Statistics Registration. England and Wales 1987 (1993). London: HMSO.

Cassileth, B.R., Volckmar, D., Goodman, R.L. (1980) The effect of experience on radiation therapy patients' desire for information. *International Journal of Radiation: Oncology-Biology-Physics* **6:** 493–496.

Cohen, F., Lazarus, R.S. (1979) Coping with the stress of illness. In Stone, G.C., Cohen, F., Adler, N.E. (eds) *Health Psychology*, pp. 217–254. San Francisco: Jossey-Bass.

Crosson, K. (1984) Cancer Patient Education. *Health Education Quarterly* **10** Special Supplement, 19–22.

Derogatis, L.R., Morrow, G.R., Fetting, J. et al. (1983) The prevalence of psychiatric disorders among cancer patients. *Journal of the American Medical Association* **249(6):** 751–757.

Eardley, A. (1986) Expectation of recovery. *Nursing Times* **82(17):** 53–54.

Fallowfield, L.J., Baum, M., Maguire, G.P. (1986) Effects of breast conservation on psychological morbidity associated with diagnosis and treatment of early breast cancer. *British Medical Journal* **301:** 575–580.

Fitzpatrick, R. (1991) Survey of patient satisfaction. *British Medical Journal* **302:** 887–889.

Graydon, J. (1988) Factors that predict patients functioning following treatment for cancer. *International Journal of Nursing Studies* **25(2):** 117–124.

Green, L.W. (1976) Site-and-symptom related factors in secondary prevention of cancer. In Fox, B.H., Isom, R.N. (eds) *Cancer, The Behavioural Dimension*, pp. 45–61. New York: Raven Press.

Gyllenskold, K. (1982) *Breast Cancer: The Psychological Effects of the Disease and its Treatment.* London: Tavistock Publications.

Hippocrates c460–c377 BC.

Holland, J., Rowland, J., Lebovits, A., Rusalem, R. (1979) Reaction to cancer treatment: assessment of emotional response to adjuvant radiotherapy as a guide to planned intervention. *Psychiatric Clinics of North America* **2**: 347–358.

Kaasa, S., Malt, U., Steiner, H. et al. (1993) Psychological distress in cancer patients with advanced disease. *Radiotherapy and Oncology* **27**: 193–197.

Levenson, L., Hamer, R.M., Rossiter, L. (1990) Reaction of psycho-pathology in general medical inpatients to use and cost of services. *American Journal of Psychiatry* **147**: 1498–1503.

Leventhal, H., Easterling, D., Coons. H., Luchterhand, C., Love, R. (1986) Adaptation to chemotherapy treatments. In Anderson, B. (ed.), *Women with Cancer: Psychological Perspectives*, pp. 172–203. New York: Springer-Verlag.

Mackenzie, C., Young, T., Maher, J., Marks, D. (1996) A study to evaluate some of the psycho-social needs of patients undergoing radiotherapy using a questionnaire format. A survey at Mount Vernon Hospital. HRC Report. London: Middlesex University.

Maher, E.J., Mackenzie, C., Young, T., Marks, D. (1996) The use of the hospital anxiety and depression scale (HADS) and the EORTC QLQ-C30 questionnaires to screen for treatable unmet needs in patients attending routinely for radiotherapy. *Cancer Treatment Reviews* (in press).

Maguire P (1990) Can communication skills be taught? *British Journal of Hospital Medicine*. **43**: 215–216.

Meyer, T., Mark, M. (1995), Effects of psycho-social interventions with adult cancer patients. A meta-analysis of randomised experiments. *Health Psychology* **14(2)**: 101–108.

Mitchell, G.W., Glickman, S. (1977) Cancer patients' knowledge and attitudes. *Cancer* **40**: 61–66.

Morris, J., Royle, G.T. (1988) Offering patients a choice of surgery for early breast cancer: a reduction in anxiety and depression in patients and their husbands. *Social Science and Medicine* **26**: 583–585.

Morrow, C.R., Morrel, C. (1982) Behavioural treatment for the anticipatory nausea and vomiting induced by cancer chemotherapy. *New England Journal of Medicine* **307**: 1476–1480.

Mumford, E., Schlesinger, H.J., Glass, G.V. et al. (1984) A new look at evidence about reduced cost of medical utilisation following mental health treatment. *American Journal of Psychiatry* **141**: 1145–1158.

Nerenz, D.R., Love, R.R., Leventhal, N., Easterling, D.V. (1986) Psychosocial consequences of cancer chemotherapy for elderly patients. *Health Services Research* **20(6)**: 961–976.

Parson, J.A., Webster, J.H., Dowd, J.E. (1961) Evaluation of the placebo effect in the treatment of radiation sickness. *Acta Radiologica* **56**: 129–140.

Paterson, A.M. (1995) *Role Development – Towards 2000: A Survey of Role Developments in Radiography*. London: The College of Radiographers.

Peck, A., Bolland, B. (1977) Emotional reactions to radiation treatment. *Cancer* **40(1)**: 180–184.

Rainey, L.C. et al. (1985) Effects of preparatory patient education for radiation oncology patients, *Cancer* **56**: 1056–1061.

Spiegel, D. (1994) Health caring – psycho-social support for patients with cancer. *Cancer* **74**: 1453–1457.

Spiegel, D., Bloom, J.R. (1983) Group therapy and hypnosis reduce metastatic breast carcinoma pain. *Psychosomatic Medicine* **45**: 333–339.

Vachon, M., Conway, B. (1989) Needs assessment of people living with cancer in a Canadian province. In Pritchard, P. (1989) *Cancer Nursing: a Revolution in Care. Proceedings of the 5th International Conference on Cancer Nursing.* London: Macmillan Press.

Young, J. (1995) A multidisciplinary approach to cancer patients. In Paterson, A., Price, R. (eds) *Current Topics in Radiography No. 1.* pp. 264–273. London: WB Saunders.

Young, J., Maher, J. (1992) The role of a radiographer counsellor in a large centre for cancer treatment: A discussion paper based on an audit of the work of a radiographer counsellor. *Clinical Oncology* **4**: 232–235.

Zigmond, A.S., Snaith, R.P. (1983) The hospital and anxiety depression scale. *Acta Psychiatrica* **67**: 361–371.

11
The Macmillan Radiographer in Cancer Care

Catherine Meredith and Lorraine Webster

In the UK and Europe cancer is a growing problem. It is a major cause of death in industrialized society where life expectancy encompasses the maximum incidence of cancer in middle and old age. Many advances have been made in the management of cancer patients and the advent of new treatment technologies have provided healthcare professionals with sophisticated methods of dealing with the physiological aspects of the condition. However, during the process of care it is important to bear in mind the needs of the patient from a wider perspective.

A diagnosis of cancer can be devastating for both patient and carers and often induces fears of pain, prolonged aggressive treatment and a possible death sentence. This, despite the fact that current cancer treatment is more successful than ever before, pain control is better understood, side-effects are able to be ameliorated and 50% of patients survive for at least five years post diagnosis (David, 1993). Nevertheless, it has been demonstrated that the diagnosis and treatment of cancer can lead to significant psychological morbidity in up to one in three patients within the first year following diagnosis (Massie and Holland, 1990).

Although cancer patients may be cared for in a variety of settings (Department of Health, 1994) many such patients are treated in departments of radiotherapy and oncology and often have as their main point of contact members of therapeutic radiography staff. It is, alas, a feature of present day radiotherapy departments that such staff are a scarce resource and patient throughput is high. The rapid development in the complexity of the technology used in radiotherapy planning and treatment delivery and the resource implications of staffing levels have caused the time available for the individual patient to be at a premium. Therapy radiographers are aware of patients' needs (Young, 1994) but feel frustrated by the time constraints placed upon them.

This observation is reinforced by the findings of the study undertaken by Cancerlink (1994) which found that patients, especially outpatients, identified the need for a contact or a key person for support and information. This study reported that patients perceived hospital staff as being very dedicated but very busy and there was not a feeling of questions being welcomed.

The Cancerlink report coupled with the increasing awareness of staff working with cancer patients that communication is inadequate, and the fact that many psychosocial problems are unrecognized (Maguire, 1985a,b) made the creation of a specialist therapy radiographer post a timely development. The need for a staff member whose role is designated as providing information and support has become more apparent as clinical departments seek to improve the quality of service offered to patients. In addition, changing fashions and expectations of patients symbolized by the advent of the Patient's Charter (Scottish Office, 1989) demand that the psycho-social dimension of care be addressed.

In the current climate of economic caution in healthcare the creation of a new post, no matter how laudable, is difficult to accomplish. The issue of funding is of pivotal importance and alternative sources to the National Health Service have to be explored. In the area of cancer care and support the track record of the Cancer Relief Macmillan Fund in sponsoring new initiatives is commendable. It was therefore decided that support for a Macmillan radiographer for the Beatson Oncology Centre (BOC), Glasgow would be sought from this source. To this end a document was prepared outlining why there was a need for such an initiative and how the post would function. Information was garnered from radiographer colleagues across the UK as to current developments in this area and this proved to be of great benefit in forming a successful case. The Cancer Relief Macmillan Fund agreed to fund this specialist post for a three year period with the proviso that funding would be taken over by the Clinical Directorate if it could be shown to have been of benefit to patients and the department.

Although the duties of the Macmillan radiographer are multifaceted three broad areas can be identified: information provision; supportive care and counselling; and education and research.

INFORMATION

Expectations of patients in respect of information about their illness are changing. Twenty years ago only a minority of patients were told they had cancer. There is now a trend to tell many patients, if not all, the precise nature of their illness. The Patient's Charter (1989) supports this and informs patients that 'you are entitled to be involved so far as is reasonably practical in making decisions about your own care, and wherever possible given choices'. However, patients' abilities to make informed choice is dependent upon the amount of information they possess. Beyond some basic facts, the majority of surveys find that the extent of knowledge about cancer in the general population is limited and often erroneous (Berman and Wandesman, 1990). There is therefore a need for patients to have access to information in both a general and a personalized manner if they wish to be involved in discussing the management of their disease.

Information is a major factor which can potentially promote an individual's feelings of control and self-efficacy, and can enhance their performance of self-care as well as sense of psychological well-being. Cancer patients are more likely to adapt to their regimes and to perform self care actions to control their symptoms when they feel they have a sense of direction and control over their health (Given and Given, 1984). By providing information and facilitating participation and choice, health care professionals can foster perceptions of control within patients. A review by Wilson-Barnett and Oborne (1983) of over twenty studies concerned with evaluating patient teaching concludes that patients do gain from and do appreciate information and feel that possession of information about their illness increases their feeling of control.

It could be argued that the duty of information disclosure should rest with medically qualified personnel, and indeed at time of initial diagnosis they are the individuals in possession of the relevant details. However, there can be little doubt that room for improvement exists. This is well illustrated by the Castejon study (1993) which found that 47% of cancer patients reported that no information had been given about the handling of their disease although the majority desired such information. Patients have indicated their dissatisfaction with this situation and the recent Report of the Health Service Ombudsman (1994) highlights particular instances where patients and their families have received less than satisfactory service. The report recommends training of hospital personnel in dealing with sensitive and disturbing diagnoses; it illustrates the importance of information and also how the relaying of it to patients requires sensitivity and well developed communication skills. This leads to the next of the broad areas to be tackled by the Macmillan radiographer.

SUPPORT AND COUNSELLING

It has been demonstrated that provision of an oncology counselling service for patients and their families ameliorates fears and anxieties about therapy (Fallowfield, 1990). The results of this study are supported by the work of Young and Maher (1992) who identified improvements in patients' anxiety levels when a radiographer counsellor was in post and improvements in their feelings of being in control. In addition, informal discussions with staff suggest that access to and use of a counselling service reduced the level of staff stress. Researchers in the growing field of psycho-oncology (Maguire, 1985a,b) advocate psychological support at an early stage in the patient's cancer journey in order to reduce the incidence of psychological morbidity at a later date.

A person centred approach to the counselling process is offered to patients at the BOC with the underlying philosophy of patient empowerment and self-help. The Macmillan radiographer's task is to enable the person to make contact with his or her inner resources. In this, the aim is to help patients accept or re-accept responsibility for their lives thus ultimately facilitating the patient in achieving an optimal quality of life.

It is recognized that the level of training in counselling skills may determine ability to recognize the degree of psychological morbidity experienced by a patient and will also determine interventions undertaken. It is of concern that many healthcare professionals working as counsellors in the demanding field of oncology have no recognizable counselling qualifications (Davis and Fallowfield, 1991). Care was taken to avoid this situation in the appointment of the Macmillan radiographer in the Beatson Oncology Centre. In order to ensure this, specific qualifications in counselling were sought in the appointed candidate and a continuing development of those skills through the educational process was incorporated into the duties of the post.

EDUCATION AND RESEARCH

The importance of education and research cannot be over-emphasized. The development of this post was initiated by a member of staff at the associated academic centre and collaboration as the post develops is ongoing. The post holder has an honorary lectureship at the academic centre and is a useful resource in student education. It became apparent in the early days of the post that many healthcare professionals working in peripheral clinics feeding the regional cancer centre had a very limited knowledge of radiotherapy and indicated that there is a demand for education from the Macmillan radiographer from a wider audience than therapy radiographers. The growing recognition of the importance of good interpersonal skills and the setting up of staff workshops to work on basic counselling skills is also a developing area.

The research component encompasses some formalized studies and an ongoing evaluation of the service provided. Information on the profile of patients requiring support and an evaluation of its effectiveness is a major component of the research undertakings. In addition, a study of the information needs of patients and how best to meet them is being undertaken in collaboration with medical and radiography colleagues. This study is being funded by the Cancer Research Campaign and although as yet unpublished has provided an interesting insight into this area.

There can be little doubt of the need for high quality education and research; the future of the radiography profession depends upon it. In the development of new roles research is essential. 'If the profession fails to recognize this, its role will not develop and other professions may demonstrate that they are better equipped to accept new roles' (Paterson, 1994).

DEVELOPMENT OF THE ROLE

As can be expected with any new initiative much effort and perseverance is required. It is important that the individual appointed to such a post is committed to its development, for at times it has proved an uphill struggle. It must be borne in mind that healthcare professionals do not work in isolation but function as members of a team. In the Beatson Oncology Centre a whole network of individuals exist who are perceived as front line patient support staff (Figure 11.1).

There is a wide perception held by the general public, and to a large extent the medical profession, that individuals involved in this type of work should be members of the nursing profession. Although initially an abreaction to this school of thought may appear quite justifiable it is indeed fact that such situations do exist. The challenge to radiographers is to alter this perception through good logical argument

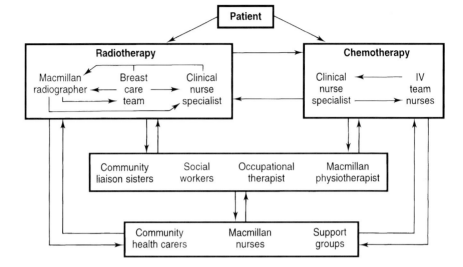

Figure 11.1 Beatson Oncology Centre system of supportive care.
(IV, Intravenous team who administer chemotherapy within day care unit.)

underpinned by research and in the process demonstrate the potential radiographers have to contribute in a wider sense to patient treatment and care.

Pursuing this policy is proving to be of positive benefit as far as the Glasgow experience is concerned. To date the Macmillan radiographer has been appointed coordinator of patient information in this centre which treats in excess of six thousand patients annually. Patient referrals are increasing from radiographers, nursing and medical colleagues and many interesting avenues for research are opening up.

THE WAY FORWARD

There is now ample evidence to justify the appointment of a support information radiographer in all radiotherapy departments. The issue of funding should be accepted as problematic but with perseverance and a good case resources can be found. Any newly appointed radiographer to such a post should be prepared for a challenging time ahead but will surely find the job very fulfilling. They must accept that they are part of a team; being territorial does not benefit radiographers or their patients. The profession has to prove itself in areas not obviously seen as radiographic by others. It takes time for a new role to become accepted. Every clinical department has its politics and its territorial boundaries. The Macmillan radiographer can cross these and in the process promote integration whilst demonstrating what radiographers have to offer.

The importance of evaluating new initiatives and publishing the results in various forums can only help promote the profession. The luxury of being insular is no longer available to radiographers if they wish to be recognized as important members of the health care team. Boundaries are shifting, new advances are being made and radiographers must be in the vanguard of change if they are to succeed in the coming years. Education and research are pivotal as is cooperation between clinically based radiographers and those in academia. Working together will enhance future developments and, perhaps, when cancer care is discussed in future no longer will it be 'nurses, doctors and who?' (Colyer, 1994).

References

Berman, S., Wandesman, A. (1990) Fear of cancer and knowledge of cancer. *Social Science and Medicine* **31:** 81–90.

Cancerlink Palliative Care Matrix (1994) *Patients' and Relatives' Views. Palliative Cancer Care Guidelines*, January. Scottish Partnership Agency and CRAG, The Scottish Office Home and Health Department.

Castejon, J., Lopez-Roig, R., Pastor, M.A., Pico, C. (1993) Cancer patients: health information and quality of life. The 7th Conference of European Psychology Society [Abstract].

Colyer, H. (1994) Nurses, doctors and who? *Radiography Today* **November:** 44.

David, J.A. (1993) A study of the role of the rehabilitation team. *European Journal of Cancer Care* **2:** 129–133.

Davis, H., Fallowfield, L. (1991) *Counselling and Communication in Health Care*, Chapter 1, Chichester: John Wiley.

Department of Health (1994) Policy Framework for Commissioning Cancer Services. London: HMSO.

Fallowfield, L. (1990) *The Quality of Life*. London: Condor Books, Souvenir Press.

Given, B.A., Given, C.W. (1984) Creating a climate for compliance. *Cancer Nursing* April, 139–147.

Maguire, P. (1985a) Deficiencies in key interpersonal skills. In Kagan, C. (ed.) *Deficiencies in Key Interpersonal Skills in Nursing*. London: Croom Helm.

Maguire, P. (1985b) Towards more effective psychological intervention in patients with cancer. *Cancer Care: Medical Services Edition* **7(2):** 12–15.

Massie, M.J., Holland, J.C. (1990) Depression and the cancer patient. *Journal of Clinical Psychiatry* **51:** 12–17.

Paterson, A. (1994) Developing and expanding practice in radiography. *Radiography Today* **60:** 9–11.
Report of the Health Service Ombudsman 1993–1994. London: HMSO.
Scottish Office (1989) Patient's Charter. London: HMSO.
Wilson-Barnett, J., Oborne, J. (1983) Studies evaluating patient teaching: implications for practice. *International Journal of Nursing Studies* **20:** 33–34.
Young, J. (1994) Counselling and communication skills courses for radiographers. *Radiography Today* **60:** 23–24.
Young, J., Maher, J.E. (1992) The role of a radiographer counsellor in a large centre for cancer treatment: a discussion paper based on an audit of the work of a radiographer counsellor. Psychosocial Oncology Group, Mount Vernon Centre for Cancer Treatment.

12
Diagnostic Imaging and Outreach Clinical Services

David Blenkinsop

INTRODUCTION

What are outreach services? These could be defined generally as the provision of services away from a central facility. How can this definition be related to radiology? If outreach imaging services away from the hospital are considered, is it possible to work outside of the normal hospital environment? Possible answers to these questions may be found in this chapter.

Why should 'outreach' diagnostic imaging services be considered. The government White Paper *Caring for People, Community Care in the Next Decade and Beyond*, (Department of Health, 1989) which preceded the 1990 National Health Service And Community Care Act, advocates that healthcare should be provided in the primary setting wherever possible, that is in the community in which patients live. The paper further states that general practitioners (GPs) will play a key role in caring for people in the community. Many GPs are rapidly developing 'hospital services' in their practices and examples of such services include physiotherapy, podiatry and district nursing.

It is necessary to consider therefore, the practicalities of providing diagnostic imaging services in a community environment. Indeed, this may be imperative from a survival point of view. Significant effort may be required, however, to adapt to this philosophy as the past twenty years or so has been spent trying to centralize imaging services onto district general hospital sites.

DEVELOPMENT OF OUTREACH CLINICS

Since the advent of general practitioner fundholding (GPFH) practices there has been a rapid development of outreach clinics and Bailey et al. (1994), found that there had been significant take up of specialist outreach clinics. Opponents of fundholding practices argue that this contributes to a two-tier healthcare service, in which better care is obtained for some patients at the expense of others. The provision of outreach diagnostic imaging services could therefore lead to some patients being disadvantaged or, at least, being perceived to be disadvantaged. There are two positions to consider in this respect. Those patients with access to outreach imaging services could have less distance to travel and may have a more complete service on one site. But, there are also benefits to those other patients who continue to use hospital based imaging services. For these patients access to the hospital based services may be easier if the outreach patients are not, for example, occupying appointment slots or taking up car parking spaces.

The many forms that outreach diagnostic imaging services might take must be considered. What range of services should or could be provided? Is on-site radiologist involvement necessary? Who will manage the services? Who will be responsible for staffing, quality assurance, health and safety, patient care and the many other issues surrounding the provision of radiological services?

Outreach diagnostic imaging services are already provided in many community hospitals throughout the United Kingdom (UK). Some of these community hospitals provide a plain film radiography service only, while some also provide a limited ultrasound service. Yet others provide opportunity for barium studies and intravenous urography. It could also be argued that mobile breast screening units are in effect, outreach diagnostic imaging services. Radiologist attendance at outreach imaging centres varies according to the nature and quantity of procedures carried out. The types of x-ray imaging equipment in community hospitals is also varied and ranges from simple mobile x-ray units to complex fluoroscopy systems. A wide variety of ultrasound equipment may also be found in community hospitals.

Consideration must be given to the quality of care that patients will receive in outreach diagnostic imaging services. Deterioration in diagnostic image quality; increased radiation doses to patients and staff, and extended report turnaround times must not be permitted and all services provided in outreach settings should achieve similar standards to those provided in the hospital environment.

With the introduction of the National Health Service (NHS) reforms, a climate of change occurred and change is progressing rapidly. Health authorities, in their new role as purchasers of healthcare, are forcing well overdue changes into working practices to ensure that their resident populations receive the best possible healthcare, whilst maintaining high quality and value for money healthcare services (Adams, 1992). Fundholding general practitioners are also negotiating and purchasing very comprehensive healthcare packages for their patients. This changing and challenging environment is generating requests for types of service provision that would previously not have been entertained. Healthcare now operates in the world of business and 'internal trading', 'marketing strategies' and 'consumer demand' are becoming familiar terms. These are, in some ways, alien concepts to healthcare practitioners. Never before has the need to operate on a solid business footing been quite so explicit. The NHS is under severe financial constraints and managers are under pressure to investigate 'income generation' schemes.

However, before looking at any area of business development in the Health Service environment it must be stressed that the primary reason for the business is the benefit of patients. This may be stating the obvious but it is a fundamental aim which must not be forgotten.

One of the new 'business' areas which could, conceivably, satisfy many of the above criteria is that of the provision of outreach diagnostic imaging services in general practices. Such schemes have already been reported in the medical press. The benefits to all parties in such partnerships are self-evident, but are dependent on local and mutually satisfactory negotiations.

The benefits to patients are many. They would not have the transport problems often associated with hospital attendance; they would be in familiar surroundings, and, in some cases, there are likely to be fewer delays in arranging appointments. In effect, patients attending an outreach imaging centre have access to a one stop service. The benefits to GPs are also significant. They would have, at least perceptually, greater control of service delivery; their patients might also be more satisfied. The benefits to the provider of the outreach service (preferably the local NHS trust) are perhaps not so immediately apparent. The most obvious benefits are that long-term relationships with the practice would be cemented and that there would be a guaranteed income for the period negotiated.

FACILITIES

The nature of diagnostic imaging work and the equipment used imposes limitations on the range of procedures that can be carried out in GP practices. At all times during the planning stage for the introduction of outreach diagnostic x-ray imaging services, it is essential to liaise with the local Radiation Protection Adviser (RPA) in order to ensure patients will be cared for in a radiation safe environment. Assuming that the many legal and beaurocratic hurdles can be negotiated satisfactorily, the range of procedures it is proposed to carry out must be considered. Again, this will be very much dependent upon the advice given by the RPA but, generally, it is often only practical to carry out plain film x-ray examinations in GP practices.

If the range of x-ray equipment available for use in such a setting is considered it is soon evident that the overall governing factor is the design or layout of the building which will house the proposed service. Rarely will it be possible to design the accommodation required to house a small x-ray unit and associated facilities. More likely the best use of an existing building will have to be made. Probably the most cost-effective way of equipping this sort of enterprise will be to utilize a redundant unit which exists in a hospital department. However, this should not be outdated. A medium frequency, well maintained generator with floor mounted tube column and floating table top is suitable. A wall mounted chest stand, or an upright Bucky, would also be necessary and, if an acceptable service is to be provided, automatic processing must be installed. It is pointless to provide an x-ray service if it is then necessary to return to a hospital to process the radiographs. With the basic set up outlined above it is possible to provide all plain film radiography and such a set up should be considered the minimum. Any thoughts of using a high power mobile unit in place of a static unit should be quickly laid to rest unless significant and, probably, expensive changes were made to the unit, particularly to the exposure hand set. Provided all x-ray equipment is thoroughly checked by the manufacturer or the engineer responsible for maintenance, and installed in accordance with the manufacturer's instructions, there is no reason for not utilizing older equipment. Any such equipment used would, of course, be subject to regular inspection by the RPA and would need a programme of quality assurance tests performed on a regular basis, the results of which must be well documented.

With ultrasound there is much greater scope for the types of equipment which can be used. Depending on finances, the equipment possibilities range from small, relatively portable machines to highly sophisticated state of the art units and there are many arguments for and against each type of unit. Overall, however, it is possible for almost all ultrasound examinations to be carried out in a GP practice.

REPORTING OF IMAGES

The question of who will provide the written reports for both ultrasound scans and x-ray examinations must be considered. Due to the dynamic nature of ultrasound it is appropriate that the person performing the examination also reports on the examination. This is stated clearly in the guidelines for professional working practice of the United Kingdom Association of Sonographers (1995). Similarly, Meire (1995) states that the ultrasound operator *must* report the scan. These arguments are not quite so straightforward in radiography where the finished radiographs have traditionally been passed to a radiologist for interpretation and report. There is, however, a growing movement which advocate reporting of plain films by radiographers. This, surely, is the way forward for outreach imaging services and in order to provide a complete one stop service to patients, is the level of service at which to aim.

OBJECTIVES

Following the National Health Service and Community Care Act of 1990, those GPs who have opted to take on fundholder status have had the ability to purchase health care for their patients from the providers of their choice. Generally, they have chosen to purchase the majority of services from the local NHS hospital trust. More and more, however, they are moving further afield in order to secure more competitive contracts (Corney, 1994). In some cases this can be to private health care providers. It has already been stated that there is a strong shift of clinical services away from the hospital into the GP surgery. In order for hospital trusts to retain GPFH contracts for diagnostic imaging services it may be essential to follow suit and move imaging services into community and primary care settings.

Radiographers must be proactive in this scenario and must ensure that the profession retains management of these new outreach diagnostic imaging services. If, for example, the equipment, the professional staff, or both are made available to provide a plain film radiography service for general practitioners it is likely that a contract for the remaining diagnostic imaging services, for an agreed period, will be secured. It is not the purpose of this chapter to examine the contractual aspects of such arrangements as these will be driven by local needs and philosophies. However, the important aims are to provide the quality of care that the purchasers and the patients expect, and to help secure the financial future of the acute hospital trust provider. It is evident that few hospitals will be able to provide this kind of service to all of its associated GPFH practices as the loss of economies of scale is all too obvious. The greatest need is likely to be in the more rural areas where transport to the nearest hospital may be difficult.

Ultrasound services, however, may open up a whole new area of enterprise. With the array of portable units on the market it is perfectly reasonable to travel further afield and attract valuable contracts from other health care purchasers. There is already a growing number of centres in which sonographers are reporting on the scans they perform, hence the problems of providing reports in a timely manner for portable ultrasound services are eliminated. Nevertheless, the ultrasound service being offered would normally be limited to obstetric, gynaecological and general abdominal scanning, although this would almost certainly be sufficient to entice most GPs.

THE FUTURE

What does the future hold for radiographers in the realms of outreach diagnostic imaging services? Survival in the new NHS requires that services be provided in accordance with the market demands.

Legislation, such as health and safety, and ionizing radiation legislation to name just two may always constrain developments. Nevertheless, it is becoming apparent that radiographers will need to become more adaptable to encompass the constantly changing health care environment. In particular, radiographers will be required to practise and report in outreach diagnostic services, often based in GP practices.

It is not inconceivable that the profession could follow the past trends seen in Australia. It is now commonplace in Australia to find private imaging services operated by radiologists in quite small towns. Many of these are also owned and operated by radiographers. Most disturbing is the fact that an increasing number of such facilities are being set up by general practitioners. Whilst they might employ qualified radiographers, which contributes to overall employment of the profession and maintains appropriate standards of care for patients, there is a concomitant reduction in available work in (patients attending) hospital departments. It can also be assumed that private/GP enterprises along these lines operate with minimum expenditure and hence with the bare minimum of radiographers. This may appear to be a doom and gloom future. However, it is a future that is in the profession's own hands. Radiographers must take a proactive approach to the market forces prevailing and must endeavour to match supply to demand. If the purchasers of health care require mini imaging departments in their own practices then it is essential that radiographers provide such facilities. Long-term comprehensive diagnostic imaging contracts may well be secured through such a proactive and professional response to purchasers' demands.

Market research is the vehicle that will enable the type and quality of services required by customers to be determined. It is radiographers' responsibilities, as professionals, to deliver these services in the most appropriate and safe manner. Outreach diagnostic imaging services are likely to become more and more the norm especially in the larger GPFH practices. Within the next few years up to 75% of GPs will become fundholders. This represents a large proportion of an acute hospitals trusts' income. In the interests of the profession and the quality of care given to patients undergoing diagnostic imaging procedures, this income must be secured.

References

Adams, S. (1992) Purchasing for quality: the provider's view. *Quality in Health Care* **1**: 65–67.
Bailey, J., Black, M.E., Wilkin, D. (1994) Specialist outreach clinics in general practice. *British Medical Journal* 23 April, 1083–1086.
Corney, R. (1994) Experiences of first wave general practice fundholders in South East Thames Regional Health Authority. *British Journal of General Practice*, January, **44**: 34–37.
Department of Health (1989) *Caring for People, Community Care in the Next Decade and Beyond*. London: HMSO.
Meire, H.B. (1995) Defensive ultrasound scanning. *RAD Magazine*.
National Health Service and Community Care Act (1990). London: HMSO.
UKAS (1995) *Guidelines for Professional Working Practice*. United Kingdom Association of Sonographers.

13
Professional Influences and Contracting for Quality of Care

Alexander Yule

INTRODUCTION

Two of the main goals of the health care purchaser are to improve patient services and secure value for money. The past five years has seen the introduction of terms such as 'contracting', 'purchasing and providing', 'marketing' and 'resource management'; these are now everyday terms. In contrast, there are terms that are often omitted, such as 'the patient', 'care of the patient' and the 'quality of care'. In order to remedy this situation purchasers are now writing into contractual frameworks specifications for quality which recognize the rights and treatment of patients in hospitals. Specifications indicate quality aspects which are aimed at improving standards of patient care but without referring to possible significant resource implications (Donaldson, 1994).

Whilst working within available resources, providers must recognize that in a competitive market, the quality of service provided could be a significant advantage to them over alternative suppliers. It is, therefore, in the providers' interest to take a positive approach in identifying the areas of quality which are necessary for successful outcomes. Providers, however, must be aware that competition in health care may not produce improvements in resource allocation (Maynard, 1994).

In most instances, the workload of radiology departments is related to internal purchasers, such as other clinical directorates, and the workload is dependent on adequate contracts being negotiated between directorates and the main purchasers. Internal purchasers probably account for 80–85% of the business, with the remainder being contracted to general practitioner (GP) fundholders, other trusts or health and local authorities.

Equity of service provided to patients can only be achieved by meeting the requirements of each patient and it must be borne in mind that patients are now increasingly aware of the Patient's Charter (Welsh Office, 1994) and the obligations it places on providers.

An important element in the provision of quality must be the involvement in the contracting process of experts from relevant professions. Radiography, as with all other health service disciplines, can be greatly affected by the type of contract negotiated at local level and it is essential that radiographers at managerial level have both the ability and the opportunity to participate in negotiations if a service that is consistent with quality matters is to be realized.

The health service, itself, is at a cross-roads, the numbers of hospitals are becoming fewer and the service provided within them is becoming more specialized and competitive. The profession of radiography is also being expected to change and to provide increased productivity for either the same or at a lesser cost. Other professions are being dismantled with duties being devolved within hospital specialties and career structures are becoming a thing of the past.

This chapter will consider the various issues which impact on contracting and show how radiographers can become involved in both the nature and the quality of the contracts.

INTERNAL CONTRACTING

Until recently, radiology departments were financed by budgets which were formulated on an historical basis. Budgets were rarely negotiated and simply rolled on from year to year with an added element for inflation and wage rises; the radiology department in the University Hospital of Wales (UHW) certainly fell into this category. However, the method of funding departments is changing dramatically and radiology managers are now involved in setting up contracts and structures to ensure that departments are run efficiently and effectively. In the UHW case this is achieved by negotiating service level agreements with the other directorates within the trust. To be included in service level agreements are staff costs, consumable costs, equipment purchase, staff training, an element for cost improvement and, of course, quality. A baseline has first to be established on which costs are based and this must be realistic. This is a two way process and the service level agreement (SLA) has to be agreed and signed by each party.

Embedded in the SLA are quality standards and the means of monitoring them. Radiographers must recognize that if they are to maintain their position in the hospital they must provide an agreed quality of service which will be monitored regularly.

EXTERNAL CONTRACTING

The number of GP fundholders are increasing in Wales and the sixth wave of fundholders have negotiated contracts which commenced on the 1 April 1996. During the past five years, the fundholders have become more aware of their powers and are negotiating separate contracts with professions such as occupational therapy and physiotherapy. Departments of radiology have provided open access for general practitioners for a number of years but there are now more and more opportunities to sell radiography as a 'stand alone' direct access service.

Contact on a one to one basis is being made with GPs and visits to surgeries, to discuss the future services with both the general practitioners and their practice managers, enable discussions to take place regarding the volume and type of work to be undertaken, the price of that work and very importantly, the quality provided.

The volume of the service is probably the easiest to decide and initially, it is based on historical data provided to the GPs and, in most cases, the GP is content to continue with the present workload. Once a GP becomes confident with respect to fundholding, discussions can take place regarding additional services which may be provided in the future. However, it must be understood that GPs may wish to negotiate an improved service for the same price or to reduce the level of service in the future and to be consulted over the development and content of contracts (Bowling et al., 1991).

The cost to the GP and the price charged by the service department is obviously important and in the competitive market the price to the purchaser must be right and affordable. Similarly, the cost charged must enable the service department not only to cover the revenue, but also to enable the department to build up a surplus to replace equipment and improve services in the future. This is a particularly difficult problem for teaching hospitals that are providing a vast range of services and has yet to be resolved.

Other external purchasers, such as health and local authorities and other trusts, are specifying those items that are considered essential to quality and also that the performance of the provider is documented and monitored on a regular basis.

Quality in the field of health care, along with its continuous improvement, is very much in today's spotlight. It is extremely important to include quality within contracts and to monitor that quality but it is a very difficult area to specify (Øvretveit, 1994). In the discussion of the volume of the service and the price, it is necessary not to lose sight of the patient and patient care, which could easily be compromised in an effort to be competitive.

AREAS OF QUALITY

The quality of services to be provided is usually set out in separate specifications contained in the contract document. Within the specifications, providers are expected to demonstrate a commitment to total quality management, adherence to statutory responsibilities, the development of quality systems, standards, indicators and the provision of good information. These standards are based on the health strategies of the purchasers, their agendas for action, the Patient's Charter and now, of course, on several years of experience. It is important, when considering standards, that professional guidance is sought by the purchasers and this would apply particularly in radiology, where the Ionizing Radiation Regulations (1988) have to be taken into account. Other documents which are helpful are the 'Guidelines for Doctors' (Royal College of Radiologists, 1993), and 'Guidelines for the Introduction of a Quality Assurance Programme in a Diagnostic Imaging Department' (College of Radiographers, 1992).

The main concerns of referring doctors must be that patients are referred appropriately, that they need the service in question and that they are properly looked after when in the hospital. To enable this to be done in radiology departments, it is essential that the professionals undertaking the examination and the treatment of patients are fully involved in discussions regarding contracts. Where quality in radiology is concerned, it must be ensured that sufficient resources are available to enable examinations to be undertaken with the minimum risk to the patient and with the greatest probability of an accurate diagnostic outcome. This can only be guaranteed if radiographers are fully involved in every stage of the contracting process.

In the competitive situation, it may be very easy for a negotiator who does not have knowledge of radiology departments to agree to a reduction in price without appreciating the need to maintain the safety and radiation requirements of each unit. It is important therefore to develop multidisciplinary teams and structures which facilitate the development of standards and quality indicators. Included in these are: quality circle programmes, corrective action groups, problem solving groups and quality improvement groups, and it is important that each discipline in the department is involved.

The production of the Patient's Charter has certainly stimulated both providers and purchasers to become more aware of patients and relatives needs. Providers now have to outline their commitment to patient care and to waiting times and treatment. To ensure compliance with the Patient's Charter, the purchasers now monitor patient satisfaction by such methods as the systematic use of questionnaires and follow-up surveys. Contracts also require hospitals and departments to provide regular reports of all complaints received, the action taken to remedy them and also the timescale within which the action has taken place. The monitoring includes waiting list time and the time taken to be seen once the patient has arrived for appointment. Without adequate safeguards such as monitoring, the performance of departments within contracts may be driven by price rather than quality.

THE INVOLVEMENT OF THE PROFESSIONAL IN QUALITY MONITORING

Despite the fact that the monitoring of the quality of services delivered is of high priority, the resourcing of these procedures is in many cases underfunded. Very often the person who undertakes this function does so in addition to other duties and as a result it may come very low down in priorities (Benton and Smith, 1992).

The radiology manager must be committed to quality and involve all staff in the development of systems. The skills and knowledge base of the professionals in the department are required to be continually updated and an on-going quality strategy must be agreed and be seen to work. The provision of a good quality system and the cooperation of the staff ensures that quality issues relating to a particular area are specified and that the performance is documented. In setting up this strategy it is therefore important to consider other groups of people who may also be concerned about the quality of the health care services provided by the radiology department.

In a document produced by the University Hospital of Wales NHS Healthcare Trust entitled 'A Strategy for Achieving Quality Improvement' published by NHS Cymru, Wales (1994), five major groups have been identified. These are:

- patients and families or carers;
- the people who live in the community served by the hospital;
- doctors, nurses, managers and other health care professionals;
- staff who are providers of care;
- health authorities and fundholding general practices, who are the purchasers of the health care services.

Also to be considered is the University of Wales College of Medicine requirements for facilities and staff to support education and research programmes for all health care professionals.

All these groups require to be assured that the health care services provided are the optimum and any quality strategy requires to take their expectations into consideration.

It must be remembered that departments of radiology are not islands within the hospital but must consider the needs of the organization as a whole. A quality service must take into account all the professionals and staff groups involved in providing a service and what they consider to be the best way to care for the patients.

Each hospital or health care service provides a broad range of very complex services. There therefore needs to be a way of ensuring the maintenance and improvement of the quality of services. This can only be done through the involvement of all professionals.

Comprehensive sets of standards of practice have been developed by radiographers and other professional groups and these have already resulted in significant improvements in the quality of patient care and it is important that these are not forgotten but are used as part of the quality contracting processes.

Views of patients and their families, in addition to those of the purchasers, must be sought to ascertain what they believe the best service should be. It is important to remember that no matter what part is played in the process of contracting, purchasing or providing the final outcome must be the provision of patient care of the highest possible quality at the lowest possible cost (Miller, 1995).

AUDIT COMMISSION REPORT ON RADIOLOGY

The Audit Commission (1995) prepared a report entitled 'Improving your Image'; the quality of service delivered and care for the patient are highlighted as important issues. The Society and College of Radiographers participated in the preparation of the document and commented in *Radiography Today* (1995).

The Commission report highlighted several areas including the provision of a better quality service. The audit team were of the opinion that radiology contributes to the effectiveness of almost every other specialty in the hospital and also to the quality of general practice in its surrounding areas. It was felt that radiology departments may become important shop windows for hospital marketing activities and it is important that managers and professionals should give careful attention to the management of radiology services.

Emphasizing this the Commission states:

> *Delivering a high quality service demands a combination of professional expertise with effective management and communication skills, both within the Department and in contact with service users.*

The Commission's report identified several tasks for radiology managers to provide a better quality service and made several recommendations as follows:

- make one individual responsible for managing the department's approach to quality improvement;
- through the individual responsible ensure that patients' and doctors' concerns are systematically addressed, as well as those identified by staff within the department;
- establish standards against which to monitor performance;
- record and follow up recommendations aimed at improving service quality.'

(Audit Commission, 1995)

These are all valid points and require to be addressed by the profession and it is important that radiographers are seen to be taking a lead in all quality performance areas. All staff should be aware of their contributions and should be encouraged to take responsibility for specific quality issues. It is important that the superintendent radiographer/manager ensures that suitable education and training is available for all staff in order that they acquire the skills and training necessary to ensure the success of the quality strategy.

CONCLUSION

This chapter has highlighted areas of quality which require to be considered in the contracting process. Setting up agreements for service contracts are not just the concern of managers in the contracts department. Each service specialty must take part in the discussions and radiology managers, in particular, must be involved in the negotiation of contracts.

Radiographers have a responsibility to ensure a quality service to patients and an obligation to secure the future of the profession. The culture of the National Health Service is changing and positive action by radiographers can greatly influence the changes.

If a quality service is to be provided to patients by continually endeavouring to meet their requirements, the achievement of high standards will become increasingly important.

The role of the radiographer is expanding rapidly and this could play an important part in future contracting negotiations. Value for money is extremely important and price coupled to quality will be an important future consideration.

The position of the radiographic profession in the health service must be maintained and improved. Radiographers must provide a quality service or the purchasers may choose to purchase the quality of care they expect from another provider. Radiographers, as providers, must therefore seek to adjust the quality of service to that required by the purchaser.

Given the current situation radiographers are in an excellent position to offer a value for money service and can look forward to the future with a great deal of confidence.

References

Audit Commission (1995) *Improving your Image*. London: HMSO.
Benton, D., Smith, M. (1992) Quality monitoring: its role in purchasing. *Nursing Standard* **6(41)**: 36–39.

Bowling, A., Jacobsen, B., Southgate, L., Formby, J. (1991) General practitioners' views on quality specifications for patients referral and care contracts. *British Medical Journal* **303**: 292–294.

College of Radiographers (1992) Guidelines for the Introduction of a Quality Assurance Programme in a Diagnostic Imaging Department. London: College of Radiographers.

Donaldson, L. (1994) Building quality into contracting and purchasing. *Quality in Health Care* **3**: 37–40.

Ionising Radiation (Protection of Persons Undergoing Medical Examination or Treatment) Regulations (1988) [POPUMET]. London: HMSO.

Maynard, A. (1994) Can competition enhance efficiency in health care? Lessons from the reform of the UK National Health Service. *Social Science and Medicine* **39(10)**: 1433–1445.

Miller, J. (1995) The birth of QUAD (1) Developing a new quality assessment document. *British Journal of Theatre Nursing* **4(10)**: 14–16.

NHS Cymru Wales (1994) *A Strategy for Achieving Quality Improvement*.

Øvretveit, J. (1994) Issues in contracting occupational therapy services. *British Journal of Occupational Therapy* **57(8)**: 315–318.

Royal College of Radiologists (1993) Making the Best Use of a Department of Clinical Radiology – Guidelines for Doctors, 2nd Edition. London: Royal College of Radiologists.

Society and College of Radiographers (1995) Audit Commission Report Urges Modernisation of X-ray Departments. *Radiography Today*, **February**, 1.

Welsh Office (1994) *Patient's Charter. A Charter for Patients in Wales*. Cardiff: Central Office of Information.

14
The Concept of Risk Management and its Applications in Radiography

Lynda Bennie

WHAT IS RISK MANAGEMENT?

The 1992 Chambers dictionary defines risk as 'the chance or possibility of suffering loss, injury, damage or failure'.

If life was without risk it would be boring and predictable. When a risk is taken and the outcome is beneficial it is felt that it was certainly worthwhile. However, every risk taken could, potentially, lead to an unwanted outcome which could result in adverse affects to people and property. As long as individuals are aware of risks, judgements can be taken about proceeding with a task because the risk is acceptable, or not proceeding because it is felt that the risk is too great. For example, some people may choose not to take an aeroplane journey for fear of an adverse incident while others calculate that the risk is minimal in comparison to other forms of transport and are quite happy to fly.

According to Dickson et al. (1991), risk is about the unfortunate things which may happen in the future. Risk management is about recognizing what these events might be, how severe they may be and how they can be controlled. Consequently, a risk management programme is designed, firstly, to identify and evaluate the risks involved in all aspects of the procedures carried out by an organization and, secondly, to institute action, relative to these risks in order to control them.

RISK IDENTIFICATION

Individuals have to take responsibility for risk identification but it is often useful to have an identified person within an organization or department who will carry out formalized risk identification procedures. It is vital, however, that this risk identification is carried out with the full cooperation of all staff members. Employees may lose motivation if incidents reported to senior staff lead to no action being taken as a result of their endeavours. The staff who work in an organization are familiar with the working practices employed and will be aware of many of the potential risks. Their opinions should, therefore, always be sought when organizational risks are being identified although it is sometimes the case that an individual who is too familiar with a situation may fail to identify a risk which an outsider would think obvious.

There are many commercial risk management companies who will carry out a risk assessment of an organization and advise on risk management. This may be expensive but can be a very worthwhile exercise, especially if risk management is a new organizational concept. The collaboration of experts in risk assessment and experts in the organization's operations will effectively enhance the risk identification and analysis programme.

To identify what risks are present the practices undertaken in the organization should be examined and the question asked, 'what potentially could go wrong?' It may be appropriate at this point to categorize the risks into areas, for example, separating risk to employees, premises and clients.

RISK ANALYSIS

Once the risks within an organization have been identified, the next stage is to analyse the impact these risks would have on the organization if any of them were to occur. What is their likely probability and frequency of occurrence and what would be their effects on the organization? This analysis can be carried out by examining both the organization concerned and similar organizations' previous losses from the identified risks, as well as thinking ahead as to what, potentially, could happen in the future. Analysing future trends and risks can never be exact but provided advice is sought from the many available sources educated guesswork is reasonably accurate.

Once the risks have been identified and analysed it is necessary to establish how these risks can be controlled. This is the final stage of a risk management programme.

RISK CONTROL

The risks in an organization have by now been identified and analysed. The third stage is to decide what action can be taken. If the risk of fire in an organization is taken as an example, it may be possible to eliminate the risk by ensuring that flammable materials are not left in corridors but are stored in appropriate fireproof containers. If the fire risk cannot be eliminated in this way, the damage caused by the fire may be reduced by the introduction of a sprinkler system; fire awareness seminars for staff; fire doors and fire extinguishers, as well as fire alarms. In this latter case, it can be seen that the organization has taken many precautions against fire damage. However, a fire could still occur with the result of amongst other things, financial costs to the organization. Whether the risk of fire is eliminated or reduced, the organization may also choose to obtain insurance which would help to cover financial losses resulting from fire.

RISK MANAGEMENT IN HEALTH CARE

A survey carried out in 1993 by the Health and Safety Executive (HSE) demonstrated that in hospitals accidental losses accounted for 5% of their total running costs.

The National Health Service (NHS), since the removal of Crown Immunity in 1991, now has to comply (as do all other organizations) with all health, safety and environmental legislation (Department of Health, DOH, 1993). It could be argued that this has given managers the impetus to ensure that the risks in their organizations are controlled adequately.

In 1990 Crown Indemnity was introduced making each NHS Trust responsible for all acts of negligence or omissions by its employees (DOH, 1993). It hardly needs to be emphasized that money spent on defending claims is money that will not be available for the treatment of patients.

In today's climate patients and staff would appear more aware of their rights and patients are more willing to complain if they are not satisfied with aspects of their care and treatment. The Patient's Charter 1991 states that NHS patients are entitled to accurate, relevant and understandable explanations of the following:

1. what is wrong,
2. what are the implications,
3. what can be done,
4. what is the treatment likely to involve.

The Charter also outlines procedural changes which should make it easier for patients to complain about any aspect of their treatment with which they are not satisfied.

In 1993 a manual was prepared by Merrett Health Risk Management Ltd for the (then) National Health Service Management Executive (NHSME) and is free to all NHS employees. The manual is recommended to all chief executives and contains information on risk management in health care with guidelines on the practical implementation of a risk management programme. Part of the recommendations were that all Trusts should identify personnel who are responsible for risk management and that a hospital risk management committee should be established.

Why is There a Need for Risk Management in Health Care?

The work carried out in a hospital invariably carries a considerable risk – from food preparation in the kitchens to the most complex surgical procedures in the operating theatres. Most of these risks will be known and accepted as part of the general working practices of the hospital. However, things can still go wrong and, unfortunately, a mistake in the clinical setting can have devastating consequences. Risk management in the hospital setting is designed therefore to reduce the costs of risk.

The costs of risk include:

- Cost in terms of human suffering. Staff, patients and patients' relatives can be adversely affected both physically and mentally if an adverse incident occurs.

- Cost in terms of the hospital's ability to provide a quality service in the perception of the purchasers of the service and the general public.
- Cost in terms of the hospital's budget as money saved from risks can be used in the treatment of patients.
- Cost in terms of the hospital's finances as litigation claims and settlements increase.

In the health service many groups of staff work as part of a well defined and circumscribed departmental team and would consider they manage their risks adequately. Risk management, if it is to succeed, should cross the inter-professional and departmental boundaries to ensure that the 'gaps' and 'overlaps' in areas of potential risk are also covered adequately.

It should be recognized, however, that there is potential for health workers to feel threatened professionally by this monitoring process which was established in industry and has been implemented by health services management. It could also be argued that it is not possible to make comparisons between private industry and an organization (the NHS) whose purpose is to improve the health of the nation.

When risk management in health care was first established it focused primarily on the financial aspect of the NHS but in recent years this has expanded to include all aspects of patient care and health and safety. It could be suggested that as long as risk management within hospitals is focusing on the financial aspects health professionals may be unwilling to become involved. However, as risk management moves into areas affecting patient care and departmental practice, health professionals must begin to participate actively as they are the experts in these areas. Risk management must have input from health care professionals as well as from managers, financial and insurance representatives to ensure a comprehensive approach and a beneficial outcome.

According to Korleski (1990) health care risk management in the United States of America (USA) emerged in response to the malpractice crisis of the mid-1970s. Korleski claims that the underlying causes of the increase in malpractice claims include technical advances in medicine which carry increased risk of injury; the deterioration of the physician–patient relationship; patients' unrealistic expectations as to the treatment outcomes, and the increased likelihood of errors due to the increased numbers of individuals involved in patient care. It is clear that health care in the USA is very different from that in the United Kingdom (UK) but in many areas the UK seems to be following the USA example and health care risk management is certainly one of these areas. Health care risk management in the USA is now well established. There are many publications in the USA on the subject and health care risk management is becoming established as a profession in its own right.

In 1980 the American Society of Healthcare Risk Management (ASHRM) was formed in response to the need for a professional association for the growing number of hospital risk managers. Further recognition of the importance of risk management was evident by the development of risk management standards by the Joint Commission on Accreditation of Healthcare Organizations (JCAHO). It may be that there are lessons to be learned from the many established health care risk management programmes developed in the USA. In particular it appears that before a risk management programme can be established the following criteria are essential:

- support from senior management,
- appointment of a risk manager,
- establishment of a risk management committee to coordinate and communicate the risk management message throughout the hospital,
- support from all personnel within the hospital,
- clear aims and objectives.

In the foreword to the Department of Health's (1993) publication *Risk Management in the NHS*, Sir Duncan Nichol, Chief Executive of the NHS Management Executive, said: 'Risk management is an important activity for all parts of the NHS. With all the changes in recent years, including the loss of Crown Immunity, it is no longer an optional extra'.

RISK MANAGEMENT IN RADIOGRAPHY

It is gratifying that to date diagnostic imaging and radiotherapy and oncology departments are not high on the list in terms of litigation claims against hospitals. However, it can take several years for a claim to be processed and documented so information available currently on the number of claims may already be out of date in terms of the present situation.

The radiographic profession has considered many of the risks involved in its practice by ensuring radiographers follow the guidelines and recommendations as described in its Code of Professional Conduct (College of Radiographers, 1994). Berlin (1994) recommends that written departmental guidelines for patient care and procedures are vital to the appropriate functioning of the department. He advises that these procedures should be developed with input from all members of the department and should reflect current practices, including incorporation of all applicable professional guidelines. It could be questioned therefore whether there is a need for a more structured approach to risk management in radiography departments.

The role of the radiographer is developing at an ever increasing rate and is being driven strongly by the internal market that now exists within the NHS (Paterson, 1995). As radiographers extend their role into new areas of expertise they may also be extending their field of work into new areas of risk. According to Berlin (1993), the risk management perspective should include technical as well as medical aspects of the service and should take into account such areas as radiographic errors; slips and falls; radiation exposure; mammography; contrast media; standards of practice; film storage; and communication.

It must be emphasized that hospitals have a duty to ensure their employees are adequately trained in the new procedures they are expected to undertake and that supervision and peer group support is available during the training period. Radiographers must ensure that in their enthusiasm to undertake new roles they do not attempt procedures until they feel confident in those procedures and have proved they are competent to proceed.

Many departments are extending radiographers' duties into areas far removed from the conventional and traditional procedures undertaken in the past; for example, administering intravenous injections and carrying out barium procedures; while those in other departments are not afforded this knowledge and experience. Radiography managers must be aware that when employing new staff familiarity with all the radiographers' duties can no longer be assumed and it may be necessary to enrol new employees on education programmes to ensure that the knowledge appropriate to a particular department is obtained.

It is often difficult in a busy department to have the resources in terms of time and money for continuing education programmes in changing techniques, protocols and risk awareness for staff, but it should be emphasized that such programmes are crucial and should be seen as a departmental priority.

Radiography education centres should extend their syllabuses to include the management of risks in undergraduate programmes as well as to provide adequate postgraduate education courses which satisfy the needs of the profession. This will mean ensuring that the programmes delivered have the flexibility to meet the changing needs of the profession thus ensuring that in the future risk management is considered part of the everyday work carried out in an X-ray or oncology department along with quality assurance and audit.

The work undertaken currently in diagnostic imaging and radiotherapy and oncology departments should be continually evaluated for both risks and improvements and, if necessary, changes in protocols and working practices should be implemented. Radiographers, together with other health professionals, should ensure that procedures and protocols are clearly documented and strictly adhered to. Staff must never become complacent when procedures appear routine as a mistake in the clinical setting can lead to catastrophic consequences for all those concerned.

CONCLUSION

The fundamental message about the management of risk is communication and education.

The successful health care risk manager must have good communication and interpersonal skills as well as an extensive knowledge of the operation of the hospital. The health care risk manager must be able to develop strong links with all hospital staff, at all levels, thus ensuring full cooperation in risk identification, analysis and control. Staff may be reluctant to notify management of mistakes, or potential adverse incidents for fear of discipline. These perceptions and fears must be allayed so a proper picture of departmental risks and incidents can be constructed. In health care a 'no blame' culture must be established with the reporting of all adverse incidents including those classified as 'near misses'. Thus an incident profile can be established and patterns of risk can be identified. However, as with all programmes of this nature, such reports must be acted upon appropriately.

Acknowledgements

The current research into Healthcare Risk Management at the Glasgow Caledonian University would not have been possible without the support of Professor John Reid, Head of Department of Risk and Financial Services and Professor Sandra Myles, Head of Department of Physiotherapy, Podiatry and Radiography.

References

Berlin, L. (1993) Liability traps and risk management in radiology – guidelines for an effective programme. *Decisions in Imaging Economics* **6(5):** Winter.

Berlin, L. (1994) *Exploring Liability Issues in Radiology.* Illinois State Medical Inter-Insurance Exchange.

Chambers Pocket Dictionary (1992) W & R Chambers.

College of Radiographers (1994) *Code of Professional Conduct.*

Department of Health (1993) *Risk Management in the NHS.* NHS Management Executive.

Dickson, G.C.A., Cassidy, D., Gordon, A.W., Wilkinson, S. (1991) *Risk Management.* The Chartered Insurance Institute.

Health and Safety Executive (1993) *The Costs of Accidents at Work.* London: HMSO.

Korleski, D. (1990) *The Emergence of a Profession, Essentials of Hospital Risk Management.* Aspen.

Paterson, A.M. (1995) *Role Development – Towards 2000. A Survey of Role Development in Radiography.* London: College of Radiographers.

Scottish Office (1991) *The Patient's Charter. A Charter for Health.* London: HMSO.

15
Magnetic Resonance Imaging and Medico-legal Work in Cerebral Palsy

Catherine Westbrook

INTRODUCTION

Cerebral palsy affects 3.7 per 1000 births. It may be defined as:

> *a disorder of posture or movement which is persistent but not necessarily unchanging and is caused by a non-progressive lesion of the brain acquired at a time of rapid brain development (Hall, 1989).*

Cerebral palsy may be accompanied by mental retardation, visual defects or epilepsy and any movement disorders are categorized as ataxic, spastic or athetoid. The causes of cerebral palsy are usually divided into prenatal, perinatal and postnatal. Perinatal causes have been traditionally associated with birth asphyxia and, not surprisingly, are increasingly common reasons for litigation (Hall, 1989).

If litigation is to be successful, proof of negligence by the obstetric team and, more importantly, causation of resultant brain damage must be proven. Negligence is often easy to establish in that most judges are swayed by an independent expert stating that they, in similar circumstances, would have managed the patient differently. In some cases this difference may only be minor.

It is much more difficult, however, to prove that any negligence directly caused the cerebral palsy and increasingly magnetic resonance imaging (MRI) is being used to assist in establishing causation. MRI is an extremely important diagnostic imaging tool of the central nervous system. Its excellent grey white matter discrimination, non-ionizing radiation and multiplanar facilities have established it as the main investigational tool in the diagnosis of brain disorders (Stark and Bradley, 1992). These advantages are especially useful in medico-legal cases in that children can be examined with great accuracy and in safety. The MRI centre at the John Radcliffe Hospital, Oxford has seen a large increase in referral of these cases by personal injury solicitors. In this chapter the role of MRI in medico-legal aspects of cerebral palsy and the radiological and ethical problems associated with these cases are discussed.

THE LEGAL PROCESS

Medico-legal medicine is well established in the United States where lawyers are employed on a 'no win–no fee' basis. Lawyers are anxious not to incur unnecessary costs on cases that they cannot win and so they often start by establishing causation since this is the stumbling block that frequently causes cases to fail. In cases of cerebral palsy two 'filtering' tests are done. Firstly, the patient and relatives undergo a series of blood tests to exclude genetic causes of the disability and secondly, an MRI examination of the brain is performed to establish whether there are lesions consistent with birth asphyxia. Litigation often only proceeds if the blood tests are normal and lesions supporting the case are present on the MRI scan. If another abnormality is found the case is severely weakened and may not proceed unless severe disability and/or obvious negligence are present.

In the United Kingdom, the medico-legal process is much less organized. Many cases do not progress due to lack of legal expertise and advice. However, a few personal injury firms are now following the United States lead and requesting MRI scans and a review of the obstetric notes by independent obstetricians as an opening gambit. The important trend is that an MRI examination is now deemed an essential step in the legal process. This may have important implications for MRI services in terms of logistics and ethics.

NEGLIGENCE

Whereas in civil criminal cases the test is 'beyond reasonable doubt', it is only necessary in civil medico-legal cases for a judge, sitting alone, to be 51% certain that both negligence and causation have been proven. The test here is, 'on the balance of probabilities'. Although cerebral palsy has been traditionally linked to birth asphyxia (Hall, 1989), epidemiological research has shown that the importance of perinatal factors in causing childhood handicap has been over-estimated (Nelson, 1988). Two major Australian studies, the Collaborative Perinatal Project of the National Institute of Neurological and Communicative Disorders and Stroke (Stanley and Watson, 1988) and a case control study of the cerebral palsy register in Western Australia (Stanley and Watson, 1985), retrospectively analysed large defined populations and attempted to estimate the proportion of children likely to have cerebral palsy as a result of intrapartum events. Both of these studies found that perinatal asphyxia is infrequently the cause of cerebral palsy (Nelson, 1988). In fact most studies estimate that over 90% of cases of cerebral palsy result from prenatal factors such as developmental abnormalities in early pregnancy (Henderson-Smart, 1991). Blair and coworkers have suggested that approximately 10% of cases of cerebral palsy are directly caused by intrapartum events (Blair and Stanley, 1988).

Birth asphyxia injury can, however, occur and its importance is great because some of it may be preventable (Nelson, 1988). Parental expectations of pregnancy and perinatal outcome are commonly very high and when things go wrong parents usually and, quite understandably, feel that there must be a reason for the child's condition and look for someone to blame. Many parents would be satisfied with a proper explanation of events but this may not be offered. Litigation may be occasionally started just to 'get to the truth of the matter'.

In addition, the long-term survival rate (defined as the second decade and beyond) of children with cerebral palsy is high and provided the child is given adequate nutrition and does not have recurrent aspiration pneumonia or episodes of status epilepticus, then there are few other life threatening insults (Harbord, 1994). As a result, parents of cerebral palsy children have to bear huge financial costs for the care and support of their child for long periods of time and therefore compensation is commonly sought through the legal system.

Birth asphyxia is difficult to define but refers in general to reduced oxygen and nutrient supply to the fetal brain (Hall, 1989). It is preferable to use the term hypoxia–ischaemia (HI) to describe this condition. HI is suspected if the following criteria are met:

- evidence of fetal distress (such as bradycardia, tachycardia, type II dips on cardiotochograph (CTG) monitor, passage of meconium stained liquor);
- severely depressed Agpar scores immediately postpartum (5 or less, 5 minutes after birth);
- depressed respiration such that the child requires significant resuscitation;
- acidosis (pH less than 7 in the first hour after birth);
- encephalopathy (fits and abnormal movements, abnormalities in feeding and sucking);
- multi-organ involvement.

Neither fetal distress, low Apgar scores nor acidosis necessarily lead to cerebral palsy (Hall, 1989). Indeed, only the last two criteria are strong evidence of significant HI (Henderson-Smart, 1991). Careful review of the obstetric notes often reveals that several of these criteria were satisfied and that serious errors of judgement by obstetric staff were made during labour or delivery. In these cases negligence is often easy to prove. Others may be more difficult but are sometimes successful even if only a degree of fault is suspected. Causation is much more complex and the role of MRI is becoming increasingly important in establishing a correlation between intrapartum events and cerebral abnormalities.

RADIOLOGICAL INFORMATION

The role of MRI in establishing causation is as follows:

- to exclude a congenital brain abnormality;
- to identify any lesions attributable to HI;
- to correlate findings with clinical conditions.

Children with congenital brain abnormalities are more likely to suffer long-lasting ill-effects from HI than those children with normal brains at term; therefore reporting any abnormalities is important. The lesions seen on MR images associated with HI are complex. Distinctive patterns have been reported (Barkovich, 1992; Rademakers, 1995; Truwit, 1992) which suggested that the type of lesions seen on MR fall into reasonably defined groups and depend on the gestational age at birth and the nature of the asphyxia (acute or chronic). In young premature infants (birth weight <1500 g), periventricular leukomalacia

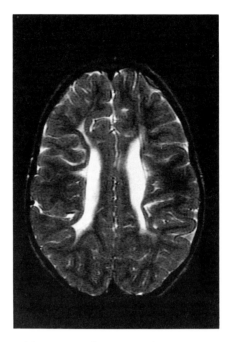

Figure 15.1 Axial fast spin echo image demonstrating periventricular leukomalacia consistent with prolonged hypoxia for several weeks before delivery.

(Figures 15.1 and 15.2) characterized by loss of periventricular white matter and cystic changes are the principal lesions (Barkovich and Truwit, 1992; Rademakers et al., 1995). In term infants, however, asphyxia causes subcortical white matter lesions and also may affect the basal nuclei (Rademakers et al., 1995; Truwit et al., 1992). Rademakers et al. report that lesions in the basal nuclei are typically located in the ventrolateral thalami and the dorsolateral part of the lentiform nucleus. They have also found that there is a pattern of left–right orientation of cortico-subcortical lesions in the pericentral region and that hippocampal atrophy may also be present. It seems that this distinctive pattern is typical in HI and may also facilitate the timing of the hypoxic ischaemic event in gestation (Rademakers et al., 1995). This may have very important implications medico-legally in that MR is often a reliable tool in establishing whether ischaemia occurred perinatally.

The reasons for the relationship between distinct hypoxic lesions and gestational age are varied. Probably the most favoured theory is one described by Azzarelli et al. (1980) and Davidson and Dobbing

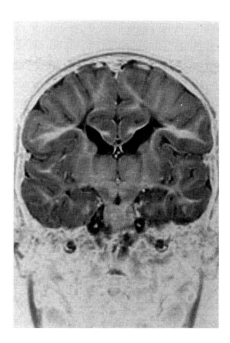

Figure 15.2 Coronal fast inversion recovery video-inverted image of the patient in Figure 15.1. The periventricular lesions appear as a dark signal.

(1966), that areas with a particularly high metabolic rate at the time of hypoxia suffer the most. This seems to correlate well with the pattern seen on MR images of lesions in the central cortico-subcortical areas of the brain in term infants that are undergoing myelination in the perinatal period. Other well-myelinated areas, such as the brain stem and cerebellum, are spared. Another theory suggests that there is an accumulation of excitotoxic substances such as glutamate because of failure of the glutamate reuptake system in hypoxia (Volpe, 1987). As a result these harmful amino acids destroy cells, especially in the hippocampus, the basal ganglia and the visual cortex. Lastly a theory suggests vascular factors are responsible where areas affected by hypoxia are situated at the edges of major vascular territories. During gestation these areas are thought to shift from periventricular regions to the subcortical areas (De Reuck, 1971). It is likely that a combination of all three theories is responsible for MR findings in any specific individual.

In order to examine these cases adequately the brain must be scanned in all three orthogonal planes using proton density, T1 and T2 weighting. In Oxford, sagittal spin echo T1 weighted images are acquired which produce good sagittal anatomy and act as localizers. They are followed by axial T2 (Figures 15.3 and 15.6) and coronal proton density and T2 fast spin echo images (Figures 15.4 and 15.5 respectively) which provide good anatomical and pathological information especially of the basal ganglia, the subcortical areas and the periventricular regions. These are acquired with a fine matrix so that high resolution is obtained. Coronal T1 gradient echo volume images are sometimes then acquired which allow for very thin contiguous slices that may be viewed in any plane. These images are especially useful for studying the hippocampal regions. The final sequence is a coronal fast inversion recovery sequence that provides excellent grey white matter contrast. This is then video-inverted using computer software so that areas with high signal on the original image appear black on the video-inversion and vice versa. This is advantageous as it alters signal appearances such that grey matter appears grey and white matter appears white (Figures 15.2 and 15.7). As the MR images are reviewed by numerous non-medical personnel during the course of legal action the inclusion of these, more easily interpretable, images have proved invaluable.

On examination of the images the radiologist must spot the lesions present and try to correlate these with the clinical findings and the medical report. An opinion can then be formed as to whether causation has been established and any factual observations and clinical conclusions are made. This is then used in evidence should the case proceed.

Table 15.1

	Sagittal SE	Oblique FSE	Coronal FSE	Coronal FIR	Coronal GE
TE (ms)	11	102	20/102	13	6
TR (ms)	460	6400	2840	5000	35
Flip (degree)	90	90	90	90	35
ETL	–	16	8	16	–
TI (ms)	–	–	–	300	–
Slice (mm)	5/2.5	6/2.5	6/2.5	6/2.5	1.5
Matrix	256×192	512×256	256×256	256×256	256×128
FOV (cm)	24	24	22	22	18
Averages	1	1	1	2	1
Options	SAT	SAT, 3/4RT	SAT, 3/4RT	SAT, 3/4RT	SAT

TE, echo time; TI, inversion time; FOV, field of view; TR, repetition time; ETL, echo train length; RT, rectangular FOV; flip, flip angle; SAT, presaturation; SE, spin echo; FSE, fast spin echo; FIR, fast inversion recovery; GE, gradient echo.

ETHICAL ISSUES

The ethics of lawyers requesting MR scans is probably one of the most difficult and contentious issues in this type of medico-legal work. It has been practice for medical practitioners to request an x-ray examination, as a request card was seen as a prescription to issue ionizing radiation and assumed that the clinician had determined that the benefits outweighed the risks. In MRI there are no known biological hazards so this criterion cannot be applied.

In addition, as the examination times are in the order of 30 minutes and as most children are either young or mentally impaired, some form of sedation is usually required to ensure that the child remains still during the examination. This naturally involves added risks to the examination and must be carefully considered. At present, the indication from British courts is that it is satisfactory to sedate these children but at Oxford it is strongly felt that general anaesthesia (GA) is the safer option. The reasons for this are as follows:

- there is intravenous access in case of an emergency;
- an anaesthetist is present should an emergency occur;
- the airway is secured during the examination;
- the time of unconsciousness is predictable;
- there is no chance of the patient waking up during the scan;
- recovery time is often faster than sedation (Westbrook, 1994);
- no conflict of interest on safety and image quality.

These benefits are, of course, weighed against the cost of purchasing magnetically safe anaesthetic equipment and an anaesthetist's and anaesthetic nurse's time (Westbrook, 1994). However, the increased safety of GA compared to sedation is an overriding factor and therefore a GA is commonly used in these cases.

Is it therefore ethical not only to scan these patients but to give them a GA at the request of a lawyer? In the opinion of the principal neuroradiologist and paediatric anaesthetist involved in these cases, if a lawyer feels that an MRI scan is in the best interests of their client and the procedure is as safe as possible, then it is acceptable for the lawyer to request an MRI scan and a GA. This opinion is not accepted widely in the radiological and anaesthetic communities and therefore medico-legal examinations of this type can only be achieved if there is cooperation of a neuroradiologist and an anaesthetist who are prepared to perform an examination at the request of non-medical personnel. It is, of course, very important to obtain informed consent from the parents and involve them in the decision-making process as it would not be acceptable for the view of the lawyer to supersede that of the parents.

CASE REPORTS

The following two cases are typical and highlight the medico-legal and radiological issues involved with these patients.

Baby 1

The mother had a history of infertility problems but eventually became pregnant using fertility drugs. Two weeks prior to the estimated date of delivery (EDD), she was admitted with high blood pressure, weight loss and a small for dates baby. She eventually went into spontaneous labour eight days prior to the EDD. When cervical dilatation had reached 3 cm, the membranes were ruptured and a scalp electrode was attached. Type II dips were noted but the reasons were thought to be due to a faulty CTG monitor. Despite this, the monitor was not replaced and the dips continued. Over 10 hours later the fetal heart was unrecordable and an emergency forceps delivery was carried out 10 minutes later. A female infant was delivered. At birth Agpar scores were 2 at 1 minute, 2 at 2 minutes and 5 at 5 minutes. The child was intubated and ventilated but then extubated 55 minutes later. Approximately 5 hours after birth, the child suffered a respiratory arrest, had a bradycardia of 40 beats per minute and was observed to have abnormal movements. She was then resuscitated but 17 hours after birth was noted to have spastic extremities. She suffered fits for 3 days and a low sodium and high urea suggesting renal impairment.

She is now 19 years old, has severe athetoid cerebral palsy, spasticity, speech and feeding problems and normal intelligence and vision. Her parents decided to take legal action, her notes were independently reviewed and she underwent an MRI examination at Oxford in February 1994.

The independent obstetrician on reviewing her case notes stated:

'The argument of the defence hinges on the interpretation of the CTG and they have no defence. It is the old story that it is not the monitoring that is at fault it is the idiots using it.'

The MRI scans demonstrated an essentially normal brain but high signal was seen in both putamina (see arrow, Figures 15.3, 15.4 and

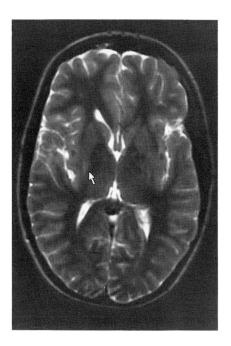

Figure 15.3 Axial fast spin echo image demonstrating high signal in the right putamen (arrow), consistent with intrapartum hypoxia.

15.5). This abnormality, despite its subtlety, is consistent with term intrapartum HI and was considered to prove that the obvious negligence during labour had indeed caused athetoid cerebral palsy. As a result, the judge awarded a very significant sum.

Baby 2

The 27 year old primigravida mother had a spontaneous rupture of her membranes eight days before the EDD. Two hours later regular contractions were established using Syntocinon. A further two hours later a scalp electrode was attached and about three hours later fetal bradycardia that lasted for five minutes was recorded by the CTG monitor. After a further half an hour, severe fetal heart decelerations followed by profound bradycardia which lasted for a full 10 minutes was noted. Fetal heart decelerations continued for a further three and half hours until birth. At birth the Agpar scores were 3 at 1 minute, 6 at 4 minutes, and 6 at 5 minutes. The child did not breathe at birth but, despite this, a paediatrician was not called for 10 minutes. On

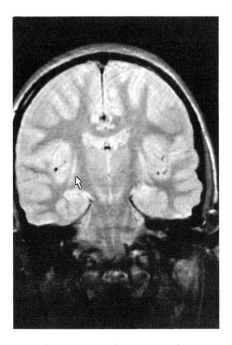

Figure 15.4 Coronal fast spin echo proton density weighted image demonstrating high signal in the right putamen (arrow), consistent with intrapartum hypoxia.

arrival, the child was intubated by the paediatrician and then extubated 45 minutes later when respiration returned. Two hours later the child suffered a fit and was noted to have a high respiratory rate and opisthotonus.

The child is now 10 years old, blind, has spastic quadriplegia, is developmentally delayed and suffers seizures that require medication. The parents began legal proceedings and her notes were independently reviewed and an MRI examination of the brain was carried out at Oxford under GA in November 1994.

The independent obstetrician on reviewing the case notes stated:

> ... the CTG trace was terminal but despite this no paediatrician was called, no action was taken to accelerate the second stage ... there was obviously a complete lack of understanding ... that this baby was seriously ill.

The MRI scans (Figures 15.6 and 15.7) revealed an atrophic brain, enlarged ventricles and severe cortical parietal damage consistent with partial, prolonged intrapartum HI in a term infant. As negligence

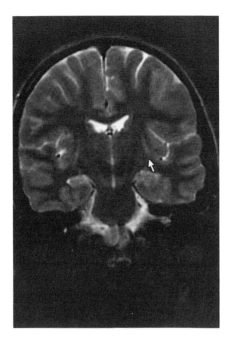

Figure 15.5 Coronal fast spin echo T2 weighted image demonstrating high signal in the left putamen (arrow), consistent with intrapartum hypoxia.

and causation had been clearly established the case was settled out of court for a significant sum.

CONCLUSION

Prior to the advent of MRI, ultrasound (US) and computed tomography (CT) were the only imaging techniques used to investigate cerebral palsy. US has limited value immediately postpartum and can only be used in the first few months. CT occasionally enabled prenatal, perinatal and postnatal asphyxia to be correlated with morphologic changes in white matter (Adsett et al., 1985). However, CT has relatively poor contrast and carries a radiation risk and is therefore not ideal in young patients. MRI has no known biological hazards and is more sensitive than CT in the detection of subtle brain malformations and mild degrees of white matter damage (Truwitt et al., 1992). For these reasons its applications in litigation cases of suspected HI causing cerebral palsy has increased.

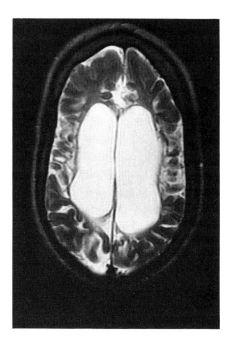

Figure 15.6 Axial fast spin echo image showing cerebral atrophy, enlarged ventricles and severe parietal cortical damage consistent with partial, prolonged intrapartum hypoxia.

Acknowledgements

I would like to gratefully acknowledge the help and expertise given by Dr Philip Anslow in the compilation of this chapter.

> COMMENT   Audrey M. Paterson, *Canterbury*

The sense of tragedy that pervades Westbrook's unemotional and factual accounts of the events relating to 'baby 1' and 'baby 2' is not mitigated by the knowledge that both received significant financial settlements. And throughout, the chapter gives rise to a deep sense of unease and of guilt – unease because it gives rise to difficult ethical and moral questions both for society and for the profession; and guilt because of the apparent impotence of the profession in relation to these questions, and especially the questions that relate directly to its practice.

Some of the questions raised are of concern not only to the profession but

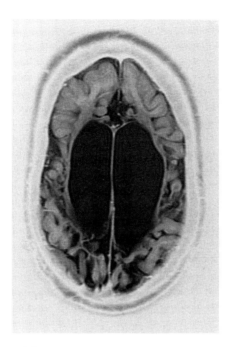

Figure 15.7 Axial fast inversion recovery video-inverted image at the same slice location as Figure 15.6.

to society as a whole. It is, for example, inexcusable that parents (or any patient) should need to begin legal proceedings simply to 'get to the truth of the matter'. The withholding of information from patients by doctors (and other health professionals) is well documented and Millman (cited in Mishler et al., 1981) describes this as an anticipatory defence mechanism against the effect of errors. But is this acceptable?

A second question for society as a whole is the financial costs that parents of children with cerebral palsy must bear and which drive many to seek compensation through the legal system. Some, as in the case of 'baby 1' and 'baby 2' are successful but, equally, some are not; and all who seek legal redress must, without doubt, bear unquantifiable emotional costs. Is it not possible for society to develop more appropriate and more equitable methods of providing financial support for these children so that they may live long and independent lives?

It is, however, the practice-related ethical matters that are of immediate concern. In essence, individuals with cerebral palsy are required to undergo a general anaesthetic of approximately 30 minutes duration and an invasive diagnostic imaging examination at the request of a lawyer. This, according to the principal neuroradiologist at Oxford, is acceptable because the procedure is as safe as possible and because the lawyer feels it is 'in the best interests of their client'. Surely, this assertion must be questioned – on what did lawyers rely prior to the advent of magnetic resonance imaging? And, surely, the opinion of a single professional from the multi-professional group of staff involved in carrying out the procedure should not determine the practice of all of those staff? Of course, and to their credit, the team at Oxford may have debated the ethics of its practice and may have the approval of its local research and ethics committee. They may also have defined clearly the limits and constraints surrounding this practice. But this is not apparent from Westbrook's work and would seem to be a serious omission.

One other aspect of Westbrook's work is cautionary. Errors in practice are evident in the case reports presented and in both cases the knowledge and competence of staff was called into question:

'... it is not the monitoring that is at fault it is the idiots using it.'
'... there was obviously complete lack of understanding ...'.

Ethically-based practice demands that staff maintain their skills and remain abreast of developments in their fields. For radiographers, this obligation is placed upon them by their Code of Professional Conduct (College of Radiographers, 1994) which states:

> Radiographers must take every reasonable opportunity to sustain and improve their knowledge and professional competence.

In the cases of 'baby 1' and 'baby 2' the consequences of professional failure in this respect were terrible.

Inevitably, the consideration of ethical issues raises more questions than it answers and no apology is made for this lack of answers. Nevertheless, there are lessons to be learned. Perhaps for radiographers these are the realization that there are significant ethical dimensions to their practice and that they must take ownership of these. By doing so, the profession will overcome its impotence in ethical matters.

For author's reply to editor's comment see p. 299.

References

Adsett, D.B., Fitz, C.R., Hill, A. (1985) Hypoxic-ischaemic cerebral injury in the term newborn: correlation of CT findings with neurologic outcome. *Developmental Medicine and Childhood Neurology* **27:** 155–160.

Azzarelli, B., Meade, P., Muller, J. (1980) Hypoxic lesions in areas of primary myelination. A distinct pattern in cerebral palsy. *Childs Brain* **7:** 132–145.

Barkovich, A.J., Truwit, C.L. (1992) Brain damage from perinatal asphyxia: correlation of MR findings with gestational age. *American Journal of Neuroradiology* **11:** 1087–1096.

Blair, E., Stanley, F.J. (April 1988) Intrapartum asphyxia: a rare cause of cerebral palsy. *Journal of Paediatrics* **112:** 515–519.

College of Radiographers (1994) *Code of Professional Conduct.* London: College of Radiographers.

Davidson, A.N., Dobbing, J. (1966) Myelination as vulnerable period in brain development. *British Medical Bulletin* **22:** 40–44.

De Reuck, J. (1971) The human periventricular arterial blood supply and anatomy of cerebral infarctions. *European Neuroradiology* **5:** 321–324.

Hall, D. (July 1989) Birth asphyxia and cerebral palsy. *British Medical Journal* **299:** 279–282.

Harbord, M.G. (1994) Medico-legal aspects of birth asphyxia: a paediatrician's perspective. *Journal of Paediatrics and Child Health* **30:** 93–95.

Henderson-Smart, D. (1991) Throwing out the baby with the fetal monitoring? *Medical Journal of Australia* **154:** 576–577.

Menkes, J.H., Curran, J. (1994) Clinical and MR correlates in children with extrapyramidal cerebral palsy. *American Journal of Neuroradiology* **15:** 451–457.

Mishler, E.G., AmaraSingham, L.R., Hauser, S.T. et al. (1981) *Social Contexts of Health, Illness and Patient Care.* Cambridge: Cambridge University Press.

Nelson, K.B. (1988) What proportion of cerebral palsy is related to birth asphyxia? *Journal of Paediatrics* **112:** 572–573.

Rademakers, R.P., van der Knapp, M.S., Verbeeten, Jr B., Barth, P.G., Valk, J. (1995) Central cortico-subcortico involvement: a distinct pattern of brain damage caused by perinatal and postnatal asphyxia in term infants. *Journal of Computer Assisted Tomography* **19(2):** 256–263.

Stanley, F.J., Watson, L.D. (1985) The methodology of a cerebral palsy register: the Western Australian experience. *Neuroepidemiology* **4:** 146–160.

Stanley, F.J., Watson, L.D. (1988) Cerebral palsy in Western Australia; Trends 1968–81. *American Journal of Obstetrics and Gynecology* **158:** 89–93.

Stark, D.D., Bradley, W.W. G. (1992) *Magnetic Resonance Imaging.* C. V. Mosby.

Truwit, C.L., Barkovich, A.J., Koch, T.K., Ferriero, D.M. (1992) Cerebral palsy: MR findings in 40 patients. *American Journal of Neuroradiology* **13:** 67–78.

Volpe, J.J. (1987) *Neurology of the Newborn,* 2nd edn. Philadelphia: WB Saunders.

Westbrook, C.A. (1994) *Handbook of MRI Technique.* Oxford: Blackwell Science.

16
Radiation Dose Limits – the UK Approach. Is it Sound?

Madge Heath

WHY ARE DOSE LIMITS NEEDED?

The International Commission on Radiological Protection (ICRP) in publication 60 (1990) stated that:

> The exposure of individuals resulting from the combination of all the relevant practices should be subject to dose limits, or to some control of risk in the case of potential exposures. These are aimed at ensuring that no individual is exposed to radiation risks that are judged to be unacceptable from these practices in any normal circumstances.

It is stated further that 'a dose limit should not be seen as a target', and 'it is the level for occupational exposure that is only just tolerable'; also that 'If the procedures of justification of practices and optimization of protection have been conducted effectively there will be few cases where limits on individual dose will have to be applied.'

THE DEVELOPMENT OF DOSE LIMITS

The fact that X-rays caused harm was realized soon after their discovery (Cooper, 1973). In 1898 the Committee on X-ray Injuries of the British Röntgen Society was formed to investigate Röntgen ray dermatitis. In 1934 a tolerance value of 0.2 Röntgen per day was adopted by the International X-ray and Radium Protection Commission. The term tolerance value seemed at that time to imply that doses below this caused little or no harm. In 1956 a report by the Medical Research Council concluded that a permissible dose of 0.3 Röntgen per week as recommended by the ICRP should continue to be accepted. In ICRP 9 (1965) the term maximum permissible dose (MPD) was used for radiation workers. The MPD for the whole body was given as 5 rem per year. The ICRP stated that the term 'has become established to describe the doses that are regarded as being the maximum that should be permitted under particular circumstances'. In ICRP 26 (1977) the term dose equivalent limit is used for both workers and the public with the annual whole body dose equivalent limit for workers remaining at 50 mSv (5 rem). In the Ionising Radiation Regulations 1985 (IRR, 1985) the term becomes dose limits and the whole body dose limit for workers remained at 50 mSv.

During the period since the discovery of radioactivity and X-rays, research has been carried out into the effects of radiation. This has led to the division of the effects of radiation into the following:

1. Deterministic effects, where there is a loss of organ function due to a large number of cells being killed or prevented from reproducing. The loss of function being greater the larger the dose, there are thresholds below which these effects do not occur.
2. Stochastic effects, which are the somatic and hereditary effects that may start from a single modified cell. The severity of the effect is not dose related.

This understanding of the effects enabled the dose limits to be set below the threshold values, thus preventing deterministic effects and reducing the probability of stochastic effects to an acceptable level.

Although the whole body dose limit did not change for over 20 years, other limits did change; the developing understanding of the effects of radiation on the lens of the eye led to the specific dose limits in ICRP 26 (Table 16.1). The ICRP came to the conclusion that 'a dose equivalent in the lens of the eye of 15 Sv would not produce opacities that would interfere with vision' and, therefore, recommended a dose equivalent limit of 0.3 Sv a year. This was revised in 1980 when the conclusion was reached that 'at this level of accumulated dose equivalent, some opacities might be produced which, while not in themselves detrimental to vision, might develop without further exposure to the point of causing deterioration of vision'. The ICRP recommended, therefore,

Table 16.1 A comparison between dose limits for occupationally-exposed workers in ICRP 26 and the Ionising Radiation Regulations 1985

Part of body	Annual dose-equivalent limit ICRP 26	Annual dose limit IRR 1985
Whole body	50 mSv (5 rem)	50 mSv
Lens of eye	0.3 Sv (reduced in 1980 to 0.15 Sv)	150 mSv
Individual organs and tissues		500 mSv

the reduction to 0.15 Sv per year. Restrictions on the time over which dose could be accumulated for women of reproductive capacity were suggested in 1965 to ensure an even division of the annual maximum permissible dose (Table 16.2). This was to ensure that an embryo did not receive more than 1 rem in the first two months of development. Further to this, the dose to the fetus once pregnancy had been declared was limited to 1 rem (ICRP 9 (1965) and Code of Practice, 1972). In IRR 1985 a similar position was held in that the dose limit for women of reproductive capacity was set at 13 mSv in any consecutive three month period and the dose limit for the abdomen of a pregnant woman was set at 10 mSv during the declared term of the pregnancy.

The United Kingdom approach to dose limits up to 1986 reflected the recommendations of the ICRP as did the practice in other countries. The MPD in the 1972 Code of Practice follows the recommendations of ICRP 9 (Table 16.2) and the dose limits in IRR85 made in response to the European Community directive follow the recommendations of ICRP 26 (Table 16.1).

Table 16.2 Maximum permissible doses for occupationally-exposed workers in the Code of Practice 1972 and the ICRP recommendations made in 1965

Part of body	Annual dose	Quarterly dose
Gonads, red bone marrow Whole body	5 rem	3 rem 1.3 rem to abdomen of women of reproductive capacity
Bone, thyroid and skin of whole body	30 rem	15 rem
Hands, forearms, feet and ankles	75 rem	40 rem
Any other single organ	15 rem	8 rem

EVIDENCE USED IN DETERMINING DOSE LIMITS

The evidence on which the dose limits are based comes from the review of evidence from many sources. The largest body of evidence comes from the Life Span Study Group of Japanese people set up following the bombing of Hiroshima and Nagasaki. This is a group of 120 321 individuals resident in Hiroshima and Nagasaki in 1950 which includes 91 228 people who were exposed at the time of the bombing. Revisions in the dosimetry of this group (DS86) (Preston and Pierce, 1988) together with a longer follow-up period have shown that the risks from radiation are higher than was previously thought. In America, the BEIR V committee (1990) reviewed the health effects of radiation in the light of the revised data and in the UK the National Radiological Protection Board (NRPB) reviewed the data for assessing stochastic effects and the effects of irradiation in utero (NRPB, 1993a).

The Japanese evidence needs to be interpreted for each population being considered as the normal incidence of cancers in Japan is different to that of Europe and America. There are various models that can be used, for example, an additive (constant excess) risk model or a relative (multiplicative) risk model where the excess incidence of cancers is expressed as a percentage of the normal underlying rates in the country concerned. The NRPB (1993a) stated that a relative risk model is currently the most appropriate model for data transfer across populations.

Much of the data on the effects of radiation on humans is gained from high dose exposures, while dose limits relate to low doses. There is still debate as to how the high dose evidence should be extrapolated to low doses.

There are various studies of populations exposed to low dose rates. However, these are not statistically powerful enough for them to be used for risk estimation. One such study is the work by the UK National Registry for Radiation Workers (NRRW) set up in 1976 to perform epidemiological studies of mortality and cancer incidence in UK radiation workers. This, so far, only covers workers in the nuclear industries. Problems always exist in interpreting dose records for radiation workers because of the lack of precision of older dosemeters (Cardis and Estève, 1991) and because different practices have been used to record dose when (a) a dosemeter has not been returned and (b) when the dose is below the threshold for the dosemeter in question, (different dosemeters having different thresholds) (Hughes and Shaw, 1991).

One of the conclusions from the first analysis of the NRRW (Kendall et al., 1992) was that 'their results taken together with a study of nuclear workers in the USA do not give any indication that the current ICRP risk estimates are significantly in error'. The ICRP (1990) noted that numerous reports appear in the literature involving the exposure of populations to low dose radiation but these studies suffer from one or more methodological problems and that more 'positive' findings tend to be reported while negative studies often are not. Concern over this publication bias was also shown by Modan (1993).

DOSE LIMITS SINCE DS 86

In November 1987 the NRPB published interim guidance on the implications of the revisions of risk estimates which show that the risks of cancer induction might be two to three times as great as originally thought. They recommended that the exposure of workers should be so controlled as to not exceed an average effective dose equivalent of 15 mSv per year over several years. In 1990 The Health and Safety Commission proposed an addition to the approved code of practice until such time as any changes are made to IRR 85. The addition would:

1. draw the attention of employers to the NRPB advice and its effect on their duty to keep doses as low as reasonably practicable.
2. provide an investigation if an individual worker's dose reached 150 mSv in 10 years.

The latter investigation is seen as a way of determining future action to ensure the exposure of the individual worker would be less in future by, for example, changing individual work patterns. This could include rotations that shared the dose out between different workers. However, Crick and Saunders (1991) stated that: 'It is a well known phenomenon that dose sharing in this way almost inevitably raises the total collective dose.' This point should be borne in mind when planning future work patterns.

In ICRP 60 (1990) the Commission concluded that 'a regular annual dose limit of 50 mSv is probably too high, and would be regarded by many as being clearly so'. They recommended a limit on effective dose of 20 mSv per year, averaged over five years (100 mSv in five years) with the proviso that the effective dose should not exceed 50 mSv in any single year. The lens of the eye retains the annual equivalent-dose limit of 150 mSv and the skin has a limit for localized exposure, to prevent deterministic effects, of 500 mSv averaged over any 1 cm^2.

The Commission recommended no special limit for women in general as they consider 'the basis for the control of the occupational exposure of women who are not pregnant is the same as that for men.' Once pregnancy has been declared they recommend a supplementary equivalent dose limit to the surface of the woman's abdomen of 2 mSv for the remainder of the pregnancy. It was suggested by Harding and Thomson (1993) that the decision to treat men and women the same is a consequence of the Gardener report, which linked excess childhood leukaemia with fathers' exposure to radiation. However, Doll (1993) stated that:

> *Clusters of leukaemia and non-Hodgkin's lymphoma in young people near some nuclear installations in the UK cannot be attributed to chance, local pollution with radioactive waste, or occupational exposure of their fathers before the children's conception.*

Little (1993) stated:

> *It therefore appears unlikely that the statistical association between recorded paternal pre-conception external radiation and the raised incidence of childhood leukaemia found in the West Cumbrian study represents a causal relationship.*

The ICRP themselves stated that:

> *If a mother, exposed prior to the declaration of pregnancy, is working under the recommended dose limits then the standard of protection for the conceptus is comparable to the standard provided for members of the general public and therefore no special limit is required for women in general.*

This implies that it is only the mother that has been considered in this instance. The guidance of the NCRP for dose limits in the USA (Hendee, 1993) is for an annual occupational effective dose limit of 50 mSv provided that the lifetime total effective dose in tens of mSv does not exceed the individual's age in years (excluding exposure to medical and background radiation).

The NRPB (1993c) made a statement on the ICRP 1990 recommendations which is broadly supportive. The NRPB recommended adopting the effective dose of 20 mSv a year as the occupational limit. They felt, however, that there is no need for the five year averaging and that there should be a constraint on optimization of any new facility much less than this dose limit. They further recommended a maximum investigative level of 15 mSv a year but go on to state that most workers' doses will not be this high so that much lower investigation limits will need to be set for most groups of workers.

Various studies have shown that the annual dose of most radiation workers in the UK is below 15 mSv per annum. Greenslade et al. (1991) examined the trends to radiation workers recorded on the central index of dose information and showed that an increasing percentage (95.8% in 1986, 96.9% in 1987, 98.1% in 1988) of workers had doses between 0–15 mSv. The workers in the higher dose bands showed a reduction of dose with time whereas those with lower doses showed no trend. Pratt and Sweeney (1991) looked at occupational exposure in the National Health Service in the north-west region; they also showed a reducing trend. In 1989 no member of the staff monitored had a dose that exceeded 10 mSv and only eight had doses between 5 and 7 mSv. Where the annual whole body dose is less than 15 mSv workers under IRR 1985 do not have to be classified. However, a reduction in the dose limit implied a reduction in the classification level.

The NRPB (1993b) recommended that workers should be divided into two groups. Group 1 would be workers who are considered likely to receive an effective dose of more than 6 mSv y^{-1} or an equivalent dose of more than three-tenths of any of the relevant dose limits for skin, extremities or lens of the eye from routine operations or accidents. Radiation dose records should be maintained throughout these workers lives and for a specified period afterwards.

Group 2 would be those workers that routinely work in a designated area but who are unlikely to receive an effective dose or an equivalent dose greater than three-tenths of any relevant dose limit.

These groups are similar to the proposed categories A and B of the European Commission (EC, 1993). Category A is the same group 1, Category B is given as those workers not classified as category A, routinely working in supervised areas or occasionally in controlled areas. The NRPB approach of not separating out the controlled or supervised areas seems to be a more commonsense way of doing the grouping.

Roberts and Temperton (1991) have analysed data for whole body doses of staff working in hospitals in the West Midlands. They showed that none of these staff needed to be classified under IRR 1985. However, they stated that if the level for classification is reduced to 5 mSv then certain occupational groups will need to be classified. This is because although only 20 people received doses of between 5 and 12 mSv in 1989 the individuals are not the same each year and therefore the whole group would need to be monitored.

One of the themes throughout ICRP 60 and NRPB (1993b) is the optimization of protection both in the design and operation stages. The ICRP included in their principles of radiological protection that:

> *should be constrained by restrictions on the doses to individuals (dose constraints), or the risks to individuals in the case of potential exposures (risk constraints), so as to limit the inequity likely to result from the inherent economic and social judgements.*

The concept of dose constraints and risk constraints were new ideas introduced with this document.

The NRPB (1993b) stated that the process by which dose constraints and investigation levels should be set for particular types of operation should include an assessment of the doses currently received during that type of operation, so that a dose constraint is set at the upper level of that which is reasonably achievable.

CONCLUSION

The NRPB guidance is more restrictive than the ICRP recommendations and the draft EC basic safety standards (1993) (Table 16.3) in

Table 16.3 Comparison of the changes proposed to the dose limits for occupational exposed persons by the ICRP, the NRPB and the EU

	ICRP 60:	NRPB 1993:	EC 1993:
Effective dose	20 mSv per year averaged over 5 years (no more than 50 mSv in any one year)	20 mSv per year	20 mSv per year averaged over 5 years (no more than 50 mSv in any one year)
Annual equivalent dose to:			
Lens of the eye	150 mSv	150 mSv	150 mSv
Skin-averaged over 1 cm^2	500 mSv	500 mSv	500 mSv
Hands and feet	500 mSv	500 mSv	500 mSv
Abdomen of a pregnant woman	2 mSv to the surface	2 mSv to the surface for photons < 100 keV	< 1 mSv to the fetus

that they consider that the 20 mSv effective dose limit should not be averaged over 10 years and that workers' doses should be reviewed quarterly to ensure they are not getting close to a dose constraint level. The NRPB do endorse the ideas of the ICRP for constraint and optimization of protection, which will ensure that workers in occupations that have lower average doses than others will have lower dose constraints thus encouraging good practice and discouraging complacency.

It is suggested therefore that the UK approach of the tighter application of dose limits and optimization of practice is a sound approach to radiation protection.

References

Beir, V. (1990) *Health Effects of Exposure to Low Levels of Ionizing Radiation*. Washington: National Academy Press.

Cardis, E., Estève, J. (1991) Uncertainties in recorded doses in the nuclear industry: Identification, quantification and implications for epidemiological studies. *Radiation Protection Dosimetry* **36(2/4):** 315–319.

Code of Practice for the Protection of Persons against Ionising Radiations arising from Medical and Dental Use (1976) London: HMSO.

Cooper, G. (1973) The development of radiation science. In Dalrymple, G.V., Gaulden, M.E., Kollmorgen, G.M., Vogel, H.H. (eds) *Medical Radiation Biology*, pp. 1–5. London: W.B. Saunders.

Crick, M.J., Saunders, P.J. (1991) The role of dose statistics in the development of occupational standards. *Radiation Protection Dosimetry* **36(2/4):** 295–298.

Doll, R. (1993) Epidemiological evidence of effects of small doses of ionising radiation with a note on the causation of clusters of childhood leukaemia. *Journal of Radiological Protection* **13(4):** 233–241.

European Commission (1993) Amended proposal for a council directive laying down the basic safety standards for the protection of the health of workers and the general public against the dangers arising from ionising radiation. *Official Journal of the European Communities* No. C 245/5.

Greenslade, E., Kendall, G.M., Fillary, K., Bines, W.P. (1991) Trends in doses to radiation workers recorded on the central index of dose information. *Radiation Protection Dosimetry* **36(2/4):** 161–165.

Harding, L.K., Thomson, W.H. (1993) Where do we stand with ICRP60? *European Journal of Nuclear Medicine* **20:** 787–791.

Health and Safety Commission (1990) Draft approved code of practice part 4. Dose limitation-restriction of exposure. Additional guidance on regulation 6 of IRR 1985. Health and Safety Commission.

Hendee, W.R. (1993) History, current status, and trends of radiation protection standards. *Medical Physics* **20(5):** 1303–1314.

Hughes, J.S., Shaw, K.B. (1991) Experience in the problems of compiling and analysing exposure data. *Radiation Protection Dosimetry* **36(2/4):** 289–293.

ICRP (1965) *Recommendations of the International Commission on Radiological Protection.* ICRP publication 9. Oxford: Pergamon Press.
ICRP (1977) *Recommendations of the International Commission on Radiological Protection.* ICRP publication 26. Oxford: Pergamon Press.
ICRP (1990) *Recommendations of the International Commission on Radiological Protection.* ICRP publication 60. Oxford: Pergamon Press.
International X-ray and Radium Protection Commission (1934) International recommendations for X-ray and radium protection. *British Journal of Radiology* **7(83):** 695–699.
Ionising Radiations Regulations (1985). London: HMSO.
Kendall, G.M., Muirhead, C.R., MacGibbon et al. (1992) First analysis of the national registry of radiation workers: occupational exposure to ionising radiation and mortality. Chilton: NRPB.
Little, M.P. (1993) A comparison of the risks of leukaemia in the offspring of the Japanese bomb survivors and those of the Sellafield workforce with those in the offspring of the Ontario and Scottish workforces. *Journal of Radiological Protection* **13(3):** 161–175.
Medical Research Council (1956) *The Hazards to Man of Nuclear and Allied Radiations.* London: HMSO.
Modan, B. (1993) Low dose radiation carcinogenesis – issues and interpretation: the 1993 G. William Morgan lecture. *Health Physics* **65(5):** 475–480.
NRPB (1993a) Estimates of late radiation risks to the UK population. Documents of the NRPB, vol. 4, no. 4. Chilton: NRPB.
NRPB (1993b) Occupational, public and medical exposure. Documents of the NRPB, vol. 4, no. 2. Chilton: NRPB.
NRPB (1993c) Board statement on the 1990 recommendations of the ICRP. Documents of the NRPB, vol. 4, no. 1. Chilton: NRPB.
NRPB (1987) Interim Guidance on the implications of recent revisions of risk estimates and the ICRP 1987 Como statement. Chilton: NRPB.
Pratt, T.A., Sweeney, J.K. (1991) Occupational exposure in the national health service. The North West Region. *Radiation Protection Dosimetry* **36(2/4):** 225–227.
Preston, D.L., Pierce, D.A. (1988) The effect of changes in dosimetry on cancer mortality risk estimates in the atomic bomb survivors. *Radiation Research* **114:** 437–466.
Roberts, P.J., Temperton, D.H. (1991) Occupational exposure in UK hospitals. *Radiation Protection Dosimetry* **36(2/4):** 229–232.

17
Gamma Camera Imaging of Positron Emitting Isotopes

A. J. Britten and J. N. Gane

INTRODUCTION

Positron emission tomography (PET) and gamma camera imaging both assess function by imaging radiopharmaceutical distribution in vivo. A PET scanner has better imaging performance than the gamma camera, but the latter has found a wider role in clinical imaging. There has recently been an increase in interest in imaging positron emitting isotopes with the gamma camera, and this chapter seeks to compare the two technologies and the clinical results achieved so far.

PET has been used extensively in a research environment, with impressive results in quantitative metabolic imaging, mainly in the fields of oncology (Strauss and Conti, 1991) and cardiology (Schwaiger and Hicks, 1991; Gould, 1991). More recently there have been initiatives aimed at applying PET imaging as part of routine radiological procedures (Fischman and Strauss, 1991; Schelbert et al., 1993), but this has proved difficult due to limited scanner availability and high procedure costs. For example, there are expected to be eight PET scanners in the UK by the end of 1995, close to the European average per country, compared to approximately 400 gamma cameras in the UK. Clinicians have wanted to access the metabolic information available by imaging the uptake of FDG (fluorine-18 fluorodeoxyglucose, an analogue of glucose) and, in the absence of a local PET scanner, some have turned to the gamma camera. Gamma camera imaging of positron emitting isotopes as a research procedure has a long history (Harper et al., 1973), but commercially available technology for positron imaging on the gamma camera has only recently become available. A number of clinical studies comparing PET and gamma camera SPECT (single photon emission computed tomography) are now being reported at meetings and in papers. There are enthusiasts for the use of gamma cameras in this area, and also detractors. As recent users of a gamma camera for FDG imaging, at a centre without a PET scanner or cyclotron on-site, the technology and clinical application of these two methodologies will be compared.

GAMMA CAMERA AND PET SCANNER TECHNOLOGY

An overview of gamma camera and PET scanner features will now be given. There are fuller desciptions of PET (Koeppe and Hutchins, 1992) and gamma cameras (Early and Sodee, 1995), and comparative work (Buddinger et al., 1977) to which the interested reader is referred. Schematic views of both systems are given in Figures 17.1 and 17.2.

Both gamma cameras and PET scanners are imaging detectors of ionizing radiation, using scintillation crystals and photomultiplier tubes (PMTs) for radiation detection, but with important differences. The main difference is that the gamma camera is essentially a large area planar imaging device which can also produce transaxial tomographs (SPECT, single photon emission computed tomography imaging), whilst the PET scanner is an inherently tomographic device. The PET scanner can only image positron emitting radioisotopes which emit two gamma rays of 511 keV each per decay, whereas the gamma camera can image radioisotopes emitting gamma rays in the range from 70 to 511 keV. Imaging of positron emitters on the gamma camera is referred to as '511 keV imaging', and not PET imaging, since PET implies use of the unique properties of the pair of gamma rays arising from positron decay. The gamma camera measures the energy of each detected gamma ray to within about 10%, and is therefore able to select or reject each gamma ray. This energy selectivity is essential to reduce the number of scattered photons detected. The energy selectivy of the gamma camera also allows the simultaneous imaging of multiple isotopes, as long as they have distinct gamma ray energies. The PET scanner has poorer energy resolution, usually accepting all gamma rays with energy above 300 keV, and therefore is poorer at rejecting scatter than the gamma camera. Fortunately the

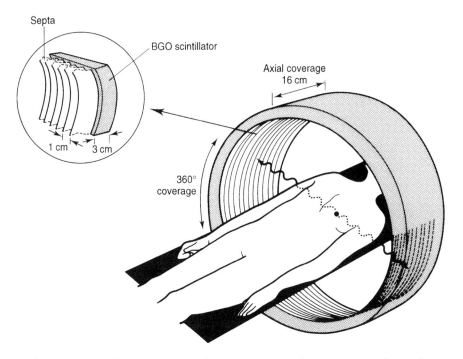

Figure 17.1 Schematic view of a PET scanner. The inset view shows the septa which reduce photons travelling oblique to transaxial planes. Both gamma rays are shown as being detected, reflecting the high stopping power of the thick BGO crystals. The axial field of view is limited.

scatter content of the detected counts can be estimated quite accurately for the PET scanner, though such corrections are difficult on the gamma camera.

Gamma camera SPECT requires the collection of data over a circular arc around the patient, and to achieve this the gamma camera head is rotated around the patient, acquiring a series of two-dimensional (2D) images which are mathematically reconstructed to produce transaxial tomographs.

The gamma camera crystal may be a circular sheet (typical diameter 40 cm) or, more commonly for modern systems, rectangular (typical dimensions 40 by 54 cm), with a thickness of 9.5 mm or occasionally 12 mm. Over 90% of 140 keV gamma rays are detected by a 9.5 mm thick crystal, but only about 10% are detected at 511 keV. The gamma camera requires the addition of a collimator in front of the crystal in order to allow only those gamma rays to pass which are perpendicular

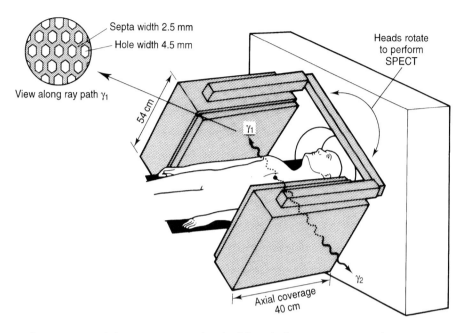

Figure 17.2 Schematic view of a dual-headed gamma camera being used to image a positron emitting isotope. The paths of the two gamma rays from one decay are shown. In this case the gamma ray γ_1 is detected, but γ_2 is undetected – illustrating the low detection probability for 511 keV gamma rays. There is a large axial field coverage, but the heads must be rotated to provide the 360 degree imaging needed for single photon emission computed tomography (SPECT). The inset figure shows the large amount of collimator lead required to reduce septal penetration of 511 keV photons.

to the crystal face. The collimator is, in essence, an array of holes in a large lead block, with a typical collimator for 511 keV imaging having hexagonal holes of 75 mm length, 3 mm lead septa and holes of 4 mm across the hexagonal flats. The collimator is the limiting feature in defining the spatial resolution and sensitivity properties of the gamma camera, with the performance of the collimator being a compromise between these two features.

The PET scanner shares many of the basic scintillation detection features of the gamma camera, but it uses the special properties of positron emitters to allow it to function without use of a collimator. The positron is the anti-particle of the electron, and when a positron is emitted in positron decay it travels a few millimetres in tissue and then combines with an electron, the mass is annihilated and converted into energy in the form of two gamma rays each of 511 keV. These two gamma rays travel along the same straight line in space, but in opposite directions. The PET scanner works by detecting both of these gamma rays, yielding the information that a positron decay took place along the line defined by joining the two detection points. Many such lines are detected and the transaxial distribution of radioisotope is mathematically calculated. There are many thousands of events detected per second, and the principle of 'coincidence detection' is used to determine which pair of events originated from the same decay. Coincidence detection looks at the time at which two gamma rays are detected. The pair of gamma rays from the positron decay start their flight at the same time, and so if two gamma rays are detected within a small time interval, typically several nanoseconds, then these two events are taken to identify a positron decay. There will be 'random coincidences' occurring when two unrelated photons are detected by chance, and techniques are used to reduce the blurring effect of these random events. Scintillation detection is used, with the scintillator arranged essentially in rings around the patient. Bismuth germanate (BGO) scintillator is commonly used, though sodium iodide (NaI) is used in some systems. Each scintillator ring defines a transaxial plane, typically around 8 mm in the axial direction, with a number of such rings to image several planes simultaneously over an axial field of view of around 16 cm (12 to 23 cm). The detection rings are separated by thin (~1 mm) lead or tungsten septa which are arranged as annular rings in front of the scintillator to reduce detection of gamma rays not originating in the plane of that ring.

In emission tomography the detected counts are affected by the scatter and absorption of gamma rays as they traverse the body towards the detectors. In SPECT the point of origin of the gamma ray is not known, and so an exact correction for attenuation cannot be made, though subsequent approximate attenuation corrections may be made (Chang, 1978). In PET, however, the actual origin of any pair of gamma rays does not need to be known, since the joint attenuation of both photons is equivalent to the total attenuation along that line through the patient. The PET scanner is therefore equipped with several removable emission rod (Ge68) sources which allow a crude form of transmission computed tomography (CT) scanning to be carried out. The resulting attenuation map is then used to perform an attenuation correction along each detected ray path, allowing the PET scanner to obtain an estimate of the absolute distribution of radioactivity within the body.

RADIOPHARMACEUTICALS

There are a range of positron emitting isotopes in use, with the commonest being those of oxygen, carbon, nitrogen, rubidium and fluorine. The half-lives of the majority of positron emitting isotopes are short, typically measured in minutes, requiring the presence of an on-site cyclotron in order to image patients. Fluorine-18 is an exception with a half-life of 110 minutes, allowing imaging to be carried out remote from a cyclotron. Both rubidium-82 and gallium-68 are available from radioactive generators, which provide a convenient on-site source or isotope. FDG (fluorine-18 fluorodeoxyglucose), an analogue of glucose, is the most widely used radiopharmaceutical, with over 200 procedures per week in Europe, followed by just over 100 studies with oxygen-15 water (European Association of Nuclear Medicine, 1995). FDG is taken up by cells at a rate proportional to glucose utilization, is phosphorylated and then essentially trapped in the cell. Regional FDG uptake therefore reflects regional glucose metabolism (Maddahi, 1994).

A wide range of isotopes and radiopharmaceuticals are used for gamma camera imaging. Technetium-99m is still the most commonly used radioisotope, with gamma camera systems performing well in imaging its 140 keV gamma ray, and with the availability of the molybdenum-technetium generator allowing a ready source of isotope on-site for radiopharmaceutical preparation. Iodine-131 at 360 keV is the commonest high energy isotope used, though strontium-85 at 514 keV has been successfully imaged (Blake et al., 1988).

COMPARATIVE IMAGING PERFORMANCE OF GAMMA CAMERA AND PET SYSTEMS

There is a wide range of system performances for both types of equipment, and typical values from modern high performance equipment in clinical use will be taken to underline essential performance differences. The introduction of low-cost PET scanners with lower performance, and high performance gamma cameras targetted at positron imaging, will be discussed in the section on recent developments.

GAMMA RAY COLLIMATION: EFFECTS ON PERFORMANCE

Gamma ray collimation is the major limiting factor for gamma camera imaging, whilst the septa of PET scanners do not present such major limitations since they only absorb rays travelling obliquely to transaxial planes. A gamma camera collimator for 511 keV imaging has lead septa of between 2.5 and 3.5 mm thick to minimize septal penetration, with a hexagonal hole size of about 3 to 4 mm and a hole length of about 75 mm. Such a collimator is heavy, in excess of 140 kg. It also has a large area of lead obstructing gamma ray passage to the crystal and is, therefore, of low detection sensitivity. The coarse hole pattern is seen on the final planar image though not on the transaxial SPECT images due to blurring in the reconstruction process. Even at these dimensions there is significant septal penetration of gamma rays, with 30% or more of detected gamma rays being oblique rays which have penetrated the lead septa (Britten et al., 1995). This septal penetration adds a wide tail to the image of a point source, reducing contrast and making quantitation difficult, and will probably be the single biggest limitation to gamma camera imaging at 511 keV. These factors make it essential to use a collimator specifically designed for imaging at 511 keV.

Current work on removing the collimator limitations is underway, with imaging without septa on PET scanners ('3D imaging') and true PET imaging without collimators being developed on gamma cameras. These developments are discussed below.

DETECTION SENSITIVITY

Good nuclear medicine imaging studies depend critically on a high detection sensitivity to obtain images with low statistical noise. Detection sensitivity depends upon the number of gamma rays able to reach the detector (geometrical acceptance), and the efficiency with which the rays are detected in the scintillator. The PET scanner has an axial extent of typically 16 cm, whilst the gamma camera is 40 cm axially, but the full 360 degree acceptance of the PET scintillator rings exceeds the limited angular coverage of the gamma camera. In addition, the PET scanner accepts gamma rays from all possible lines within a transaxial plane, whereas the gamma camera only accepts those gamma rays parallel to the radius of the gamma camera head rotation, due to the collimator acceptance.

Detector sensitivity is strongly dependent upon thickness of the crystal. The 9.5 mm thick sodium iodide crystal of the typical gamma camera has a detection sensitivity at 511 keV of only about 10%, whilst the 30 mm BGO PET crystal detects approximately 94% of 511 keV photons. Since two gamma rays must be detected simultaneously in the PET scanner the final detection sensitivity is the single photon detection probability squared, about 88%.

Modern gamma camera systems may have two or three imaging heads as a simple means of increasing the detection sensitivity.

Further consideration must be made of the effect of attenution in the patient. Patient attenuation has a much greater effect in PET than in SPECT, since for PET to detect both 511 keV gamma rays both must escape from the body. This factor reduces the 511 keV photons available for detection in PET by a factor of about 2 compared to SPECT (Buddinger et al., 1977).

Final tomographic performance figures are that a typical PET scanner has a detection sensitivity of approximately 200 000 cps/ (μCi ml^{-1}), whilst a dual headed gamma camera yields about 5500 cps/(μCi ml^{-1}), both for a uniformly F-18 filled 18 cm diameter phantom. There is little information available regarding relative sensitivity in patient imaging, but a factor of 10 to 30 in favour of the PET scanner is expected. One paper (Martin et al., 1995) reports a relative sensitivity advantage of 8 for the PET scanner in patients, whilst Chen et al. (1995a) report a factor of 35 in a uniformly filled phantom. Further sensitivity comparisons in patients are required. In clinical practice this difference in sensitivity reveals itself as a difference in scan time and image noise. For example, 30 minute SPECT myocardial imaging with FDG has been reported (Martin et al., 1995), whilst the corresponding PET scan can be acquired in under 10 minutes. This ratio of acquisition times does not directly reflect underlying system capabilities, probably due to the fact that 30 minute SPECT images are acceptable, whilst reducing PET scan time to less than 10 minutes is of little value and so images of better statistical quality are acquired. The PET sensitivity advantage becomes essential, however, in low count studies, and gives the PET scanner the ability to detect lower contrast low-uptake tumours, for example, as well as the ability to perform dynamic tomography at a rapid frame rate. Dynamic scans have the potential of providing unique information on tissue function, allowing uptake rates and extraction fractions to be calculated. This information is not available from SPECT studies since it appears that 30 minutes is a practical minimum for FDG cardiac studies, and 45 minutes may be more appropriate for oncology studies. The biodistribution of radioisotopes must be approximately constant over these SPECT imaging times, otherwise artefacts will occur in the reconstructed slices, and whilst this requirement is reasonably well satisfied from 45 minutes after

FDG injection it may be a limitation for other radiopharmaceuticals, and clearly rules out imaging of short half-life positron emitters.

SPATIAL RESOLUTION

The spatial resolution in the transaxial slice is defined as the full width at half maximum (FWHM) of the profile through a line source parallel to the axis of the system. The axial spatial resolution is the corresponding FWHM of a point source lying in a transaxial plane. The transaxial and axial spatial resolutions are usually different, and also vary with the distance of the source from the axis. Spatial resolution drops rapidly with distance of the source from the gamma camera collimator, and care should be taken to consider spatial resolutions which can be achieved in clinical imaging. Measurements in pseudo-anthropomorphic phantoms show that a FWHM of about 20 mm is achievable in the heart for SPECT imaging of FDG (Britten et al., 1995). This is to be compared to a PET transaxial spatial resolution of about 7 mm. Poorer spatial resolution affects the contrast of the final image, more so for the detection of reduced uptake than increased uptake, and so the threshold contrast detectability will be lower for SPECT than PET. Furthermore, the FWHM is only part of the information about spatial resolution. The high degree of septal penetration in the gamma camera collimator (30% or more for a 511 keV collimator) leads to long tails on the line spread function. These long tails reduce the image contrast and are a major problem in trying to quantify regional uptake in a SPECT image at 511 keV. One approach to improve the spatial resolution of SPECT is to use fan beam collimators, and initial results with a 511 keV fan beam collimator for the brain are good, with a spatial resolution of 7.9 mm (Patton et al., 1995).

COUNT RATE CAPABILITY

The PET scanner has a high count rate capability, usually of the order of several million coincidences per second. The count rate is high for clinical PET imaging due to the large detector sensitivity, and the high acceptance of scattered gamma rays. The gamma camera has a relatively poor count rate performance, typically of the order of 50 000 counts per second as a practical maximum. This is not a problem for clinical gamma camera studies, since most clinical doses give count rates of less than 4000 counts per second. Current attempts to use the gamma camera without the collimator as a coincidence PET detector require special hardware to increase the gamma camera's limited count rate capability (see below).

MULTIPLE ISOTOPE STUDIES

The energy selectivity of the gamma camera means that multiple isotopes can be imaged simultaneously, and FDG and 99mTcMIBI have been used to obtain simultaneous images of myocardial perfusion and blood flow within one 30 minute SPECT study (Delbeke et al., 1995; Sandler et al., 1995). The PET scanner can only image positron emitters, and therefore identification of different isotopes is not possible. The short half-life of most positron emitters is used to make up for this lack of simultaneous imaging, since it allows sequential studies to be performed, such as 13-N ammonia blood flow studies followed by 18-FDG glucose metabolism studies.

QUANTITATIVE IMAGING

Gamma ray scatter and absorption, and limited spatial resolution are sources of error in radioisotope imaging. A fundamental advantage of the PET scanner is its ability to measure and correct for non-uniform body attenuation and therefore measure radioactive concentration in terms of MBq ml^{-1} in vivo. The problems of attenuation, poorer spatial resolution, septal penetration and scatter radiation make quantitation difficult for SPECT, but there is steady progress towards achieving better SPECT quantitation (Rosenthal et al., 1995). It is interesting to note that recent developments in gamma camera technology are aimed at increasing its quantitative capabilities, whereas for the PET scanner quantitative accuracy is being reduced in clinical imaging in order to shorten the lengthy scanning times.

RECENT DEVELOPMENTS IN GAMMA CAMERA TECHNOLOGY

Multiple heads on gamma cameras are now becoming more prevalent, thereby increasing the detection sensitivity. Dual-headed cameras are more common than triple-headed cameras, since they are better adapted to a range of planar imaging which forms a large part of routine clinical imaging, and also on cost grounds.

Gamma ray attenuation has a direct impact on cardiac imaging where the inferior wall of the myocardium may appear abnormal due to increased attenuation relative to the anterior wall. Systems which allow transmission computerized tomography to be performed during emission imaging are now being clinically evaluated (Stewart et al., 1995). In principle this technology could be applied to 511 keV SPECT imaging in order to improve quantitative accuracy, though this remains to be proven. The greater penetration of 511 keV photons may produce less attenuation artefact than at lower energies (van Balen et al., 1995), but attenuation correction is still required for accurate quantitative studies.

Detection of scattered gamma rays is a major source of contrast loss in SPECT imaging. The facility is now available to estimate and subtract scatter through the use of multiple energy window acquisitions which sample the compton-scattered spectrum and allow an estimate of the scatter fraction within the photopeak. This estimate improves image contrast, but evaluation of its accuracy is awaited.

Removal of the collimator from the gamma camera system allows, in principle, a multi-headed gamma camera to use the same coincidence detection technology as a PET scanner. Such devices have been proposed since 1975 (Muehllehner, 1975), but it is only recent advances in digital electronics which have made this practical. Overall the intrinsic detection efficiency is low, of the order of 1%, due to the low stopping power (10%) of each 9.5 mm scintillator and the need to detect both gamma rays from each decay. The large crystal area presents problems with the detection of activity outside the field of view. The number of true events is about 1% of the total counts detected, and so to obtain 4000 events per second the gamma camera must be capable of processing 400 000 cps, and special electronics are needed to deal with this high count rate.

Initial evaluation of a prototype has shown a spatial resolution of the order of 5 mm for brain imaging (Muehllehner et al., 1995), and further clinical evaluation is being undertaken. If successful, this would allow such a hybrid device to carry out positron imaging in coincidence mode without the collimators, replacing the collimators to carry out single photon imaging. A hybrid scanner capable of brain imaging with PET and single photon isotopes was available in 1981 (Kanno et al., 1981), but is no longer commercially available.

RECENT DEVELOPMENTS IN PET SCANNER TECHNOLOGY

PET procedures have been lengthy, and the need to achieve a higher throughput has seen developments aimed at reducing the overall imaging time. This includes the use of high sensitivity 3D imaging, and improvements in transmission scanning.

3D imaging involves the retraction of the attenuating septa which define the ring geometry. In this mode pairs of gamma rays are detected which lie at an angle to transaxial planes. These gamma rays would have been attenuated by the lead septa when working in the traditional transaxial mode. The sensitivity of the PET scanner in 3D mode is increased, probably by a factor of 3 or 4, though comparison between the two modes is not straightforward due to increases in scatter acceptance in 3D imaging (DeGrado et al., 1994; Kinahan et al., 1995). Reconstruction of the 3D data into the 2D transaxial slices is more complex than in the traditional mode, but the increase in detection sensitivity may be valuable in clinical oncology imaging when large areas of the body are imaged in order to screen for previously unsuspected metastases.

Transmission imaging has traditionally required scanning before the patient is injected with isotope, and this lengthens the procedure as well as causing potential misalignment problems if the patient moves. Systems have now been developed which allow transmission scanning after injection, and developments are in progress to allow simultaneous acquisition of transmission and emission images. These developments allow a greater throughput at the sacrifice of some quantitative accuracy, with 3D imaging having a higher scatter fraction and with the newer transmission scanning modes experiencing cross-contamination of emission and transmission data (DeGrado et al., 1994; Kinahan et al., 1995; Meikle et al., 1995). This loss in absolute accuracy may not be of consequence for clinical imaging since in clinical diagnosis visual assessment may be adequate, or even superior to quantitation (Stollfuss et al., 1995; Lowe et al., 1994; Knuuti et al., 1993).

The majority of centres cannot perform these newer modes of transmission scanning and so many are evaluating whole body imaging without attenuation correction. The uncorrected images are distorted, particularly in areas of superficial uptake, but initial reports suggest that the diagnostic information may be similar to the more accurate attenuation corrected data (Hoh et al., 1993).

The capital cost of a PET scanner has been over £1 000 000 and specialized low cost scanners have been developed over many years (Cherry et al., 1989). Recently the main suppliers have developed reduced cost scanners, about £600 000, to meet commercial constraints, whilst maintaining adequate diagnostic imaging performance. This may be achieved by reducing the amount of scintillator used with partial rings and then rotating the scintillator, much like an X-ray CT scanner. Initial evaluation of such low cost systems shows good performance (Myers et al., 1995). Also, sodium iodide (NaI) may be used as a scintillator instead of BGO, with the advantage of reduced cost, with an approximately 50% loss of detection efficiency due to the reduced stopping power of NaI.

CLINICAL FDG GAMMA CAMERA IMAGING – THE LEFT VENTRICULAR MYOCARDIUM

Gamma cameras have been used to image the left ventricular myocardium, with early planar studies of nitrogen-13 ammonia (Harper et al., 1973) and rubidium-81 (Berman et al., 1975), but with the majority using fluorine-18 FDG. (Britten et al., 1995; Martin et al., 1995; Delbeke et al., 1995; Hoflin et al., 1989; Williams et al., 1992; Bax et al., 1993; Van Lingren et al., 1992; Drane et al., 1994; Stoll et al., 1994; Burt et al., 1995; Kalff et al., 1995). The myocardium is an obvious first choice for FDG gamma camera imaging, since PET FDG scanning has been shown to be the most accurate technique to identify hibernating myocardium (Di Carli et al., 1994), and since the motion of the heart probably makes very high resolution imaging unnecessary. Hibernating myocardium has low perfusion and contactile dysfunction (Rahimtoola, 1987), but will restore its function if revascularized by coronary artery bypass grafting (CABG) or angioplasty (PTCA). Other gamma camera imaging for myocardial perfusion (thallium or technetium sestamibi, TcMIBI) tend to erroneously identify hibernating myocardium as infarcted (Tamaki et al., 1991; Dilsizian et al., 1994). FDG is an ideal radiopharmaceutical to identify hibernation since the poorly perfused muscle primarily utilizes glucose instead of fatty acid, and so there is high or maintained uptake of the glucose analogue FDG corresponding to areas of reduced perfusion. Experience with planar gamma camera FDG imaging has been recently reported. Huitink et al. (1995) report a study with a mean of four years clinical follow-up of 69 patients following myocardial infarction, with planar thallium and FDG studies. Increased uptake of FDG in the regions with reduced blood flow was found to be highly predictive of future adverse cardiac events. Kalff et al. (1995) compared FDG PET with FDG planar gamma camera imaging. There was overall good agreement between the two methodologies, but differences occur in comparing the defects rated as 'small' or 'moderate' by PET scanning: of 38 such PET defects gamma camera imaging rated 9 as normal. The age of the gamma camera (1974) and the known limitations of planar imaging may contribute to such discrepancies. The majority of current studies are with emission tomography, SPECT imaging. Direct comparison of myocardial FDG SPECT and PET scanning in the same patients has been reported by several authors. Taking PET as the gold standard, Burt et al. (1995) report that SPECT identified 11 out of 14 'PET viable' segments, and with agreement in 45 out of 47 non-viable myocardial segments. A similar agreement was found in 25 patients by Chen et al. (1995), with 137 out of 140 segments in agreement between FDG PET and SPECT. Burt et al. (1995) also carried out rest thallium imaging as the best known conventional gamma camera technique for identifying hibernating myocardium. Of 20 patients, 7 had areas of viability on FDG PET (8 on FDG SPECT) not identified with resting thallium imaging. This suggests that FDG SPECT may be significantly better than the thallium rest injection procedure, though further studies are required with more detailed follow-up to identify hibernating myocardium. Initial results of such detailed studies have recently been given from Amsterdam. Cornel et al. (1995) report initial results in 20 patients comparing FDG SPECT with dobutamine echocardiography (DE), concluding that DE and FDG SPECT have similar predictive value for the outcome of coronary artery bypass grafting (CABG). Work by Bax et al. (1995) in 5 patients consisted of FDG SPECT with PET, thallium perfusion and 2D echocardiography prior to and following CABG. The results suggest some differences between FDG PET and SPECT, but further patient numbers are required.

The detection of hibernating myocardium requires imaging of myocardial perfusion as well as the FDG glucose metabolism image, and separate thallium perfusion studies have mainly been used. The gamma camera's ability to discriminate gamma ray energies has been used to obtain simultaneous images of perfusion with 99m-Tc-sestamibi (TcMIBI) at 140 keV and glucose metabolism with FDG at 511 keV (Britten et al., 1995; Delbeke et al., 1995; Sandler et al., 1995; Stoll et al., 1994). Simultaneous imaging has advantages in that the perfusion and glucose metabolism images are exactly aligned, and this facilitates comparison, as well as allowing perfusion and glucose uptake to be imaged within 30 minutes. This is to be compared to approximately 60 minutes or more required for PET perfusion/FDG imaging of the heart. Phantom data and these initial patient studies suggest that high quality diagnostic images of glucose metabolism and perfusion can be obtained simultaneously, despite the known problems of 'cross-talk' of 511 keV photons scattered into the 140 keV energy window, differing relative attenuation between the two isotopes, and the spatial resolution of the order of 20 mm.

CLINICAL FDG GAMMA CAMERA IMAGING – ONCOLOGY AND BRAIN

FDG gamma camera studies in oncology have been reported (Martin et al., 1995; Van Lingren et al., 1992; Drane et al., 1994; Macfarlane et al., 1995), with fewer non-oncology brain studies (Martin et al., 1992; Schaefer et al., 1995). Drane et al. (1994) reported FDG SPECT followed by 30 minute whole body planar imaging in 47 patients with tumours. In 43 of these the primary tumours and confirmed lymph node metastases were visualized on gamma camera imaging. The three false negatives were a low grade lymphoma and two low-grade gliomas. In five of the 47 patients a primary or metastatic tumour, confirmed at subsequent surgery, was seen on gamma camera imaging but not by previous radiology. Comparison of FDG SPECT and FDG PET in the same patient has been reported (Martin et al., 1995; Schaefer et al., 1995). In a group of 8 patients being assessed for malignant tumours Martin found agreement between FDG PET and FDG SPECT in 7, with two lesions of less than 15 mm being missed by FDG SPECT in one patient and only faintly visualized by FDG PET. MacFarlane et al.'s (1995) experience with a three-headed gamma camera is less encouraging, since in 13 patients 22 tumours were identified by FDG PET but only 11 were seen with FDG SPECT. Four of the 11 tumours detected on SPECT were 20 mm or less, whilst 9 of the 11 tumours missed were 20 mm or less. FDG SPECT detection of small tumours will be strongly dependent on absolute amount of uptake in the tumour and the tumour-to-background ratio, due to the poorer spatial resolution and detection sensitivity than for PET. Absolute uptake values for the tumours were not quoted, and further studies are required which define detectability thresholds in terms of absolute tumour uptake and tumour-to-background ratios.

Brain imaging for evaluation of intracerebral lesions has been approached with caution since it is expected that the poor spatial resolution, especially for low uptake regions, will reduce the ability of FDG SPECT. Four patients are reported by Martin et al. (1995), with the overall conclusion that lesions of more than 20 mm may be visualized, but that FDG PET imaging shows significant advantages. The same group has proceeded to investigate the application of a high resolution fan-beam collimator for brain imaging, and initial evaluation (Patton et al., 1995) shows a spatial resolution of 7.9 mm, and good phantom images. Phantom studies and one patient (Schaefer et al., 1995) imaged at a spatial resolution of about 11 mm indicate that hypermetabolic lesions above 20 mm may be visualized.

DISCUSSION

The comparison of imaging performance clearly shows that the PET scanner is the optimal device for imaging positron emitters. However, even though the gamma camera is not optimized for 511 keV imaging, modern systems have shown themselves to be capable of producing images of diagnostic quality. Gamma camera FDG imaging of the heart has been reported to identify hibernating myocardium, and certain centres are so confident that this is now a routine clinical procedure since, though PET images are of superior quality to their gamma camera equivalents, the diagnostic accuracy remains similar (Delbeke et al., 1995; Drane et al., 1994). Direct comparison of FDG SPECT and FDG PET for the detection of hibernating myocardium is valuable to evaluate the technique, but further comparison of FDG SPECT with other widely available techniques is important. In particular, the study of Burt et al. (1995) has shown that FDG SPECT is better than rest thallium imaging for the identification of viable myocardium, with over one-third of patients with 'infarct' identified by thallium having viable myocardium identified with FDG PET and FDG SPECT. If these results are reproduced elsewhere then the use of FDG SPECT would have a significant impact on patients being selected for revascularization procedures. FDG SPECT has also been shown to be comparable with dobutamine echocardiography (Cornel et al., 1995) for identifying hibernating myocardium.

Most clinicians do not have PET imaging available, whereas the availability of gamma camera SPECT is high and gamma cameras can now be purchased with the 511 keV imaging option or even upgraded in the field, with costs of this option in the range £20 000 to £50 000. However, it must be recognized that the wider application of the PET scanner in the fields of oncology and neurology has yet to be duplicated by the gamma camera. Indeed, the superior spatial resolution, low count and low contrast detection abilities of the PET scanner may prove to be essential in order to provide the high diagnostic accuracy required in these important areas.

There are insufficient clinical data so far, with preliminary series reporting sensitivities between 50% (Hamberg et al., 1994) and 91% (Drane et al., 1994) for tumour detection when compared to PET scanning. Thorough comparison with other widely used imaging techniques is also required, such as x-ray CT and magnetic resonance imaging (MRI), since though the gamma camera may perform poorly against the PET scanner it may yet outperform other anatomically based imaging modalities and may therefore still find a clinical role.

It appears unlikely that there will be dramatic performance improvements in conventional gamma camera imaging at 511 keV with collimators, but the developments of coincidence counting on the gamma camera offers the promise of high spatial resolution and improved detection sensitivity. Though initial work on this hybrid gamma camera/PET system appears promising it will be some time before the limiting effects of poor count rate capability, and the influence of the 'out-of-field' activity, will be known. The first clinical evaluation of coincidence PET on a gamma camera is expected during 1996. If the known problems of gamma camera coincidence detection can be overcome satisfactorily for clinical imaging then the complete integration and widespread availability of PET imaging as part of clinical nuclear medicine departments may become a reality. This will still be limited to fluorine-18 radiopharmaceuticals or generator produced isotopes, unless there is an on-site cyclotron, but in clinical practice to date it is FDG which has shown the greatest value in cardiac and oncological imaging.

Will the availability of such hybrid PET/SPECT scanners signal the end of the dedicated PET scanner? This appears unlikely since the dedicated PET scanner will still be able to provide quantitation through its ability to calculate attenuation maps, and higher count rate capability and detection efficiency to allow dynamic studies. This is particularly so in the field of oncology where quantitation from dynamic studies may yet be of significant value in assessing tumour metabolism. This may be particularly true in the emerging area of monitoring tumour response to chemotherapy, in which there are sound theoretical advantages to performing dynamic imaging and measuring tumour uptake rate constants (Hamberg et al., 1994; Fischman and Alpert, 1993). In addition, work is in progress on other more specific positron labelled tumour markers, such as C-11 amino acids to identify the high protein synthesis rate in tumours (Kole et al.,

1995), which may offer more specific tumour uptake than that shown by FDG. C-11 imaging requires the higher count rate performance of the PET scanner.

Is the current interest in FDG gamma camera imaging a threat to PET centres? The authors believe that gamma cameras can provide clinically valuable information which is not currently available due to the currently limited distribution and high cost of PET scanning. The wider availability of FDG imaging will raise the awareness of clinicians to the unique and valuable information provided by FDG imaging, and to the use of other PET radiopharmaceuticals. This awareness may be used to generate demand for high quality FDG imaging, and the newer generation of low cost PET scanners supplied with FDG at a low cost from remote cyclotrons may well be economically justifiable. Specialist centres with cyclotrons will be able to provide a wider range of PET procedures, and their currently high cyclotron running costs may be recovered by the widespread commercial supply of FDG to remote gamma camera and PET scanner centres. Such a model is already developing in Germany where FDG doses are available now for less than £100, and several PET scanners operate without a cyclotron on site.

The overall justification for FDG imaging lies in the need to correctly target expensive and life threatening operations, both for cardiac revascularization and tumour resection. PET is moving into the clinical arena, and gamma camera imaging of FDG should assist that process.

CONCLUSION

There has been a recent growth in the gamma camera imaging of positron emitting isotopes, due in part to technological improvements in gamma cameras. Modern gamma cameras have good performance across a wide energy range, and multiple headed cameras provide enhanced detection sensitivity. The instrument of choice to detect positrons is the PET scanner which has been described and compared to the gamma camera. The limited availability of PET scanners has hindered the benefits of PET being made widely available, and gamma cameras are being evaluated as a more widely available imaging device to fill this need. The reported work on cardiac imaging has been reviewed and it may be concluded that gamma camera imaging of the glucose analogue fluorodeoxyglucose (FDG) produces diagnostic information similar to that from a PET study. The limited data from gamma camera imaging of FDG in oncology is less convincing, and further work is required. Advances in gamma camera and PET scanner technology have been discussed. The effects of reduced cost PET scanners coupled with shorter PET scan times due to technological advances, and the wider awareness of FDG imaging made possible by the use of gamma cameras, are expected to promote the wider clinical use of PET scanning.

References

Bax, J.J., Visser, F.C., van Lingren, A. et al. (1993) Fluorine-18-fluorodeoxyglucose and SPECT to detect viable myocardium after recent infarction [abstract]. *European Journal of Nuclear Medicine* **20**: 841.

Bax, J.J., Visser, F.C., Veening, M.A. et al. (1995) Functional recovery after revascularisation: comparison between FDG-PET and FDG-SPECT. *European Journal of Nuclear Medicine* **22**: 798.

Berman, D.S., Salel, A.F., De Nardo, G.L., Mason, D.T. (1975) Non-invasive detection of regional myocardial ischaemia using rubidium-81 and the scintillation camera. *Circulation* **52**: 619–626.

Blake, G.M., Zivanovic, M.A., Blaquiere, R.M. et al. (1988) Strontium-89 therapy: measurement of absorbed dose to skeletal metastases. *Journal of Nuclear Medicine* **29**: 549–557.

Britten, A.J., Gane, J.N., Gill, J.S. et al. (1995) Evaluation of dual isotope 511 keV and 140 keV SPECT imaging on dual headed gamma camera (abstract). *European Journal of Nuclear Medicine* **22**: 809.

Buddinger, T.F., Derenzo, S.E., Gullberg, G.T. et al. (1977) Emission computed tomography with single photon and positron annihilation photon emitters. *Journal of Computer Assisted Tomography* **1**: 131–145.

Burt, R.W., Perkins, O.W., Oppenheimer, B.E. et al. (1995) Direct comparison of Fluorine-18-FDG SPECT, Fluorine-18-FDG PET and rest Thallium-201 SPECT for detection of myocardial viability. *Journal of Nuclear Medicine* **36**: 176–179.

Chang, L.T. (1978) A method for attenuation correction in radionuclide computed tomography. *IEEE Transactions on Nuclear Science* **NS-25**: 638–643.

Chen, E.O., MacIntyre, W.J., Go, R.T. et al. (1995a) Differences in imaging characteristics of PET and 511 keV SPECT: implications for myocardial viability studies. *European Journal of Nuclear Medicine* **22**: 809.

Chen, E.O., MacIntyre, W.J., Go, R.T. et al. (1995b) FDG SPECT imaging of viable myocardial segments: a comparison with FDG PET technique (abstract). *Journal of Nuclear Medicine* **36**: 3P.

Cherry, S.R., Marsden, P.K., Ott, R.J. et al. (1989) Image quantification with a large area multi-wire proportional chamber positron camera (MUP-PET). *European Journal of Nuclear Medicine* **15**: 694–700.

Cornel, J.H., Bax, J.J., Firetti, P.M. et al. (1995) FDG/thallium SPECT versus dobutamine echocardiography in the prediction of improved left ventricular wall motion after surgical revascularisation. *European Journal of Nuclear Medicine* **22:** 797.

DeGrado, T.R., Turkington, T.G., Williams, J.J. et al. (1994) Performance characteristics of a whole-body PET scanner. *Journal of Nuclear Medicine* **35:** 1398–1406.

Delbeke, D., Videlfsky, S., Patton, J.A. et al. (1995) Rest myocardial perfusion/metabolism imaging using simultaneous dual-isotope acquisition SPECT with technetium-99m-MIBI/fluorine-18-FDG. *Journal of Nuclear Medicine* **36:** 2110–2119.

Di Carli, M.F., Davidson, M., Little, R. et al. (1994) Value of metabolic imaging with positron emission tomography for evaluating prognosis in patients with coronary artery disease and left ventricular dysfunction. *American Journal of Cardiology* **20:** 527–533.

Dilsizian, V., Arrighi, J.A., Diodati, J.G. et al. (1994) Myocardial viability in patients with chronic coronary artery disease: comparison of 99mTc-Sestamibi with Thallium reinjection and 18F-Fluorodeoxyglucose. *Circulation* **89:** 578–587.

Drane, W.E., Abott, F.D., Nicole, M.W. et al. (1994) Technnology for FDG-SPECT with a relatively inexpensive gamma camera-work-in-progress *Radiology* **191:** 461–465.

Early, P.J., Sodee, D.B. (eds) (1995) *Principles and Practice of Nuclear Medicine*, 2nd edn. Mosby Year Book.

European Association of Nuclear Medicine (1995) *Task Group Report: Positron Emitters*. European Association of Nuclear Medicine, Keizersgracht 782, 1017 EC Amsterdam, The Netherlands.

Fischman, A.J., Strauss, H.W. (1991) Clinical PET – a modest proposal. *Journal of Nuclear Medicine* **32:** 2351–2355.

Fischman, A.J., Alpert, N.M. (1993) FDG-PET in oncology: there's more to it than looking at the pictures. *Journal of Nuclear Medicine* **34:** 6–11.

Gould, K.L. (1991) PET perfusion imaging and nuclear cardiology. *Journal of Nuclear Medicine* **32:** 579–606.

Hamberg, L.M., Hunter, G.J., Alpert, N.M. et al. (1994) The dose uptake ratio as an index of glucose metabolism: useful parameter or oversimplification? *Journal of Nuclear Medicine* **35:** 1308–1312.

Harper, P.V., Schwartz, J., Beck, R.N. et al. (1973) Clinical myocardial imaging with nitrogen-13 ammonia. *Radiology* **108:** 613–617.

Hoflin, F., Lederman, H., Noelpp, U. et al. (1989) Routine 18F-2deoxy-2-fluoro-D-glucose (18-FDG) myocardial tomography using a normal large field of view gamma camera *Angiology* **40:** 1058–1064.

Hoh, C.K., Hawkins, R.A., Glaspy, J.A. et al. (1993) Cancer detection with whole-body PET using 2-18F-Fluoro-2-Deoxy-D-Glucose. *Journal of Computer Assisted Tomography* **17:** 582–589.

Huitink, J.M., Visser, F.C., Bax, J.J. et al. (1995) Clinical outcome of patients with myocardial infarction after viability studies with planar 18F-fluorodeoxyglucose imaging, (abstract). *European Journal of Nuclear Medicine* **22:** 777.

Kalff, V., Berlangieri, S.U., van Every, B. et al. (1995) Is planar thallium-201/fluorine-18 fluorodeoxyglucose imaging a reasonable clinical alternative to positron emission tomographic myocardial viability scanning? *European Journal of Nuclear Medicine* **22**: 625–632.

Kanno, I., Uemura, K., Miura, S. et al. (1981) Headtome: a hybrid emission tomograph for single photon and positron emission imaging of the brain. *Journal of Computer Assisted Tomography* **5**: 216–226.

Kinahan, P.E., Jadali, F., Sashin, D. et al. (1995) A comparison of 2D and 3D PET imaging. *Journal of Nuclear Medicine* **36**: 7P.

Knuuti, M.J., Nuutila, P., Ruotsalainen, U. et al. (1993) The value of quantitative analysis of glucose utilisation in detection of myocardial viability by PET. *Journal of Nuclear Medicine* **34**: 2068–2075.

Koeppe, R.A., Hutchins, G.D. (1992) Instrumentation for positron emission tomography: tomographs and data processing and display systems. *Seminars in Nuclear Medicine* **22**: 162–181.

Kole, A.C., Nieweg, O.E., Pruim, J. et al. (1995) Early clinical experience with the tumour tracer L-[1-C-11]tyrosine in positron emission tomography. *European Journal of Nuclear Medicine* **22**: 836.

Lowe, V.J., Hoffman, J.M., DeLong, D.M. et al. (1994) Semiquantitative and visual analysis of FDG-PET images in pulmonary abnormalities. *Journal of Nuclear Medicine* **35**: 1771–1776.

Macfarlane, D.J., Cotton, L., Ackermann, R.J. et al. (1995) Triple-head SPECT with 2-[fluorine-18]fluoro-2-deoxy-D-glucose (FDG): Initial evaluation in oncology and comparison with FDG PET. *Radiology* **194**: 425–429.

Maddahi, J. (1994) *Journal of Nuclear Medicine* **35**: 707–715.

Martin, W.H., Delbeke, D., Hendrix, B. et al. (1995) FDG SPECT: Correlation with FDG-PET. *Journal of Nuclear Medicine* **36**: 988–995.

Meikle, S.R., Bailey, D.L., Hooper, P.K. et al. (1995) Simultaneous emission and transmission measurements for attenuation correction in whole-body PET. *Journal of Nuclear Medicine* **36**: 1680–1688.

Muehllehner, G. (1975) Positron camera with extended counting rate capability. *Journal of Nuclear Medicine* **16**: 653–657.

Muehllehner, G., Geagan, M., Countryman, P., Nellmann, P. (1995) SPECT scanner with PET coincidence capability (abstract). *Journal of Nuclear Medicine* **36**: 71P.

Myers, M.J., Bailey, D.L., Bloomfield, P.M. et al. (1995) ECAT ART – a low cost BGO PET camera using rotating detectors. *Journal of Nuclear Medicine* **36**: 70P.

Patton, J.A., Ohana, I., Weinfeld, Z. et al. (1995) DISA FDG/MIBI SPECT with parallel hole and fan beam collimators. *Journal of Nuclear Medicine* **36**: 167P.

Rahimtoola, S.H. (1989) The hibernating myocardium. *American Heart Journal*. **117**: 211–221.

Rosenthal, M.S., Cullom, J., Hawkins, W. et al. (1995) Quantitative SPECT Imaging: A Review and recommendations by the Focus Committee of the Society of Nuclear Medicine Computer and Instrumentation Council. *Journal of Nuclear Medicine* **36**: 1489–1513.

Sandler, M.P., Videlefsky, S., Delbeke, D. et al. (1995) Evaluation of myocardial ischaemia using a rest metabolism/stress perfusion protocol with fluorine-18 deoxyglucose/technetium-99m MIBI and dual isotope simultaneous-acquisition single-photon emission computed tomography. *Journal of the American College of Cardiologists* **26**: 870–878.

Schaefer, A., Reiche, W., Kirsch, C.M., Piepgras, U. (1995) Single Photon emission computed tomography of brain tumours using 18-Fluorodeoxyglucose: methodological aspects and first results (abstract). *European Journal of Nuclear Medicine* **22**: 809.

Schelbert, H., Bonow, R.O., Geltman, E. et al. (1993) Position statement: clinical use of cardiac positron emission tomography. Position paper of the cardiovascular council of the society of Nuclear Medicine. *Journal of Nuclear Medicine* **34**: 1385–1388.

Schwaiger, M., Hicks, R. (1991) The clinical role of metabolic imaging of the heart by positron emission tomography. *Journal of Nuclear Medicine* **32**: 565–578.

Stewart, R., Ponto, R., Dickinson, C. et al. (1995) Comparison of the simultaneous transmission-emission protocol (STEP) with parallel hole SPECT for tc-99m Sestamibi perfusion imaging: validation with PET. *Journal of Nuclear Medicine* **36**: 169P.

Stoll, H.-P., Hellwig, N., Alexander, C. et al. (1994) Myocardial metabolic imaging by means of fluorine-18 deoxyglucose/technetium-99m sestamibi dual-isotope single photon emission tomography. *European Journal of Nuclear Medicine* **21**: 1085–1093.

Stollfuss, J.C., Glatting, G., Friess, H. et al. (1995) 2-(Fluorine-18)-fluoro-2-deoxy-D-glucose PET in the detection of pancreatic cancer: value of quantitative image interpretation. *Radiology* **195**: 339–344.

Strauss, L.G., Conti, P.S. (1991) The applications of PET in clinical oncology. *Journal of Nuclear Medicine* **32**: 623–648.

Tamaki, N., Ohtani, H., Yamashita, K. et al. (1991) Metabolic activity in the areas of new fill-in after thallium-201 reinjection: comparison with positron emission tomography using fluorine-18-deoxyglucose. *Journal of Nuclear Medicine* **32**: 673–678.

van Balen, S., Hoogweg, H., van Loon, M.J.A. et al. (1995) Does photon attenuation affect the interpretation of circumferential profiles when comparing Tl-201 and F18-FDG SPECT images of the myocardium? *European Journal of Nuclear Medicine* **22**: 946.

Van Lingren, A., Huijgens, P.C., Visser, F.C. et al. (1992) Performance characteristics of a 511-keV colimator for imaging positron emitters with a standard gamma camera. *European Journal of Nuclear Medicine* **19**: 315–321.

Williams, K.A., Taillon, L.A., Stark, V.J. (1992) Quantitative planar imaging of glucose metabolic activity in myocardial segments with exercise thallium-201 perfusion defects in patients with myocardial infarction: comparison with late (24hr) redistribution thallium imaging for detection of reversible ischaemia. *American Heart Journal* **124**: 294–304.

18

MRI of the Breast

Joanna Linas and Elizabeth M. Warren

INTRODUCTION

The United Kingdom (UK) has the highest breast cancer mortality rate in the world but a high incidence and mortality rate is also a feature common to other Western countries. World wide, over 500 000 new cases of female breast cancer are diagnosed each year but half of these occur in Europe and North America which has less than 20% of the female world population. Breast cancer is the most common cancer amongst women of the UK and each year 26 000 women are newly diagnosed. It is estimated that one in twelve women will develop breast cancer at some time in their life, resulting in 16 000 deaths per annum and making it the leading cause of cancer deaths in women between the ages of 35 and 54. Very few cases occur in women in their teens or early twenties but the annual incidence rises from over a thousand in the 35–39 age group to over 3000 in the 60–64 age group (Cancer Research Campaign, 1991).

In this chapter, after reviewing the present treatment regimes and imaging methods of breast disease, the development of magnetic resonance imaging (MRI) as a diagnostic tool in this field and the use of intravenous contrast agent are explored. Current MRI techniques are described along with an evaluation of the clinical applications. Finally, the limitations and the likely future potential of MRI in breast disease diagnosis are discussed.

TREATMENT

The best chance of survival from breast cancer is achieved if treatment provides optimal local control and accurate staging of individuals with advanced regional multifocal or distant disease (Harms et al., 1992). The group of women who die from breast cancer have distant occult metastases at the time of local treatment or develop metastases after inadequate treatment of local or regional disease (Harms et al., 1992). Many patients presenting with a clinical diagnosis of localized breast cancer have micrometastatic disease which is undetectable by currently available imaging.

Successful breast cancer treatment requires early detection of the neoplasm before it has metastasized outside the breast followed by surgical removal of the lesion. Operative techniques minimize deformity and usually conserve the breast and may be followed by radiotherapy and or chemotherapy (Harms and Flamig, 1993). In general it can be surmised that the earlier the breast cancer is diagnosed the better the prognosis. The current success level of treatment results in an average of 62% of women alive five years after breast cancer diagnosis (Harms and Flamig, 1993).

Surgical removal of the primary lesion either involves a mastectomy or breast conservation. Conservation surgery involves local excision ensuring a 1 cm margin of normal tissue is removed along with axillary surgery for staging of subsequent treatment and prognosis (Dixon, 1993). Breast conservation surgery is only suitable for patients who have a single clinical or mammographic lesion, less than 4 cm in size, showing no signs of local advancement, nodal involvement or metastases. Larger lesions of over 4 cm are treatable with conservation surgery in patients with large breasts. Mastectomy is the treatment of choice in patients with larger tumours, lesions positioned centrally in small breasts, multifocal disease or widespread lymphatic invasion (Dixon, 1993).

Whilst some advanced breast carcinomas may be diagnosed by clinical examination the rather grim statistics have stimulated the search for an imaging technique which will detect a small primary carcinoma at an early stage before metastatic spread. Such a technique may either be applied to symptomatic patients or used to screen an asymptomatic population, although the exact requirements in terms of sensitivity and specificity are somewhat different.

CURRENT IMAGING

Mammography

The first reported use of radiography for breast imaging was as early as 1913 for mastectomy specimens, however it was not until the 1930s that a method of mammography in vivo was described (Warren, 1930). Various levels of enthusiasm were demonstrated over the following twenty years with no significant advancements until 1960. A technique described by Egan using industrial film, a high milliampere (mA) and low kilovoltage (kV) resulted in a clinical trial featuring very positive results in the detection of breast carcinoma. Subsequent technological developments led to a dramatic improvement in resolution and contrast providing greater confidence in diagnosis. Inspired by the early positive findings of Egan and improved technology, a number of large randomized screening trials were undertaken during the 1960s and 1970s. The findings of reduced mortality and high level of sensitivity led to the realization of benefits associated with mammography screening (Heywang-Köbrunner, 1990). Screening trials in Europe and elsewhere have supported a survival benefit in postmenopausal women but evidence is less convincing for the younger age group. This was also corroborated by a large and very ambitious Canadian National Breast Screening study but evidence from Sweden indicates probable efficacy for both pre- and postmenopausal women.

X-ray mammography has been the most readily available and practical technique for detection of the primary lesion, although in older women 5–15% of palpable carcinomas are not demonstrated and in women younger than 50 years only 56% of malignancies are identified. The strength of mammography lies in the diagnosis of malignancy in the fatty breast and the excellent demonstration of micro calcifications, which are present in about 30–50% of early malignancies (Heywang-Köbrunner, 1990). The reasons for some lesions going undetected include sub optimal technique, observer limitations and tumour size and nature. In the dense breast, for example, the sensitivity of mammography significantly decreases, particularly in patients without microcalcification (Homer, 1985). Often the lesions that are detected are indeterminate and require a biopsy to diagnose or exclude cancer. A recent study carried out by the two major British cancer charities and the Medical Research Council into breast screening has concluded that the accuracy of a screening mammogram can be increased by 25%, if two views of each breast are performed instead of one view (Wald et al., 1995).

Other Diagnostic Techniques

Historically, imaging of the breast has centred on mammography; however, other imaging methods have been proposed over the last forty years but with varying degrees of efficacy and economic feasibility in the detection of disease.

Ultrasound has proved useful in distinguishing between solid and cystic lesions and therefore helps in reducing unnecessary biopsies (Heywang-Köbrunner, 1990). The technique is also used to confirm a palpable malignancy within mammographically dense tissue. Ultrasound is, however, not a reliable method of excluding malignancy as many lesions, particularly those that are diffuse or non-invasive, are not visible.

Histologically, malignant tumours tend to be more vascular than benign lesions and this vascularity can be identified with Doppler ultrasound (Jackson et al., 1993). Currently Doppler does not have a role in the clinical management of lesions and only tends to be used in the evaluation of solid masses (Cosgrove et al., 1990). Colour Doppler may prove to have a future in differentiating between benign and malignant solid masses but it is doubtful that it will be valuable as a screening tool for either fatty or dense breasts (Jackson et al., 1993).

Transillumination and thermography have both been explored in breast diagnostics however neither have proved to be useful due to the high level of false positive and false negative results (Heywang-Köbrunner, 1990). The cost and spatial resolution have been the limiting factors in breast computed tomography (CT), and an injection of iodinated contrast agent is required making it unsuitable for screening. Studies involving nuclear medicine techniques have also been explored, for example thallium uptake has been proved to be higher in hormone dependent breast cancer (Sluyser and Hoefnagel, 1988). Recent research with positron emission tomography shows that uptake of labelled oestrogen and glucose may occur at primary breast cancer locations and occasionally at sites of lymph node metastases (Mintun et al., 1988). Nuclear medicine studies are unlikely to be used routinely for breast cancer detection due to high expense and time requirements although further investigation is necessary to establish its role.

Although a number of methods are already available for early detection and differentiation of breast carcinoma, additional information is still desirable to increase the sensitivity in the mammographically dense breast and increase specificity, particularly in differentiating small lesions.

THE DEVELOPMENT OF BREAST MRI

The limitations of mammography coupled with the increasing concern associated with the use of ionizing radiation have sparked recent interest in MR imaging of the breast. The use of MRI in imaging the breast is not a new technique as some of the first MR images produced in 1978 were of the breast (Harms and Flamig, 1993). Early clinical trials predicted a promising future for MRI in breast cancer diagnosis. However, by the late 1980s this picture had changed revealing that MRI had little to offer over conventional imaging.

It had been thought that improved tissue characterization might be possible by evaluating the MRI properties of the different breast tissues. These properties can be assessed by evaluating T1, T2 and proton density and the signal intensities the tissues assume on different pulse sequences (Heywang-Köbrunner, 1990). Tissue differentiation based on tissue parameters proved to be more difficult than expected. This was due to the significant overlap of T1 and T2 relaxation times between benign and malignant breast tissue, making them difficult to distinguish.

The impression that breast MRI would have no future continued until the introduction of gadolinium contrast agents, where research with this agent revealed that breast carcinomas consistently enhanced. This improved the tissue differentiation making it possible to distinguish between benign and malignant tissue (Hickman et al., 1994).

The introduction of gadolinium based contrast agents improved the specificity of breast MR resulting in a renewed interest in its application. The potential of MR imaging in the evaluation of breast disease has continued to increase over the past decade in line with advances in MR technology (Hylton and Frankel, 1994). In particular the development of surface coils with increased signal to noise ratios and high resolution fast scanning techniques have contributed to the improved imaging methods available (Hylton and Frankel, 1994).

Enhanced Breast MRI

In view of the difficulties experienced in tissue differentiation of an unenhanced MRI examination the use of contrast agents in MRI of the breast is essential.

MRI contrast agent

MRI intravenous contrast agents, due to their magnetic properties, are able to accentuate the image contrast between normal and abnormal tissue by altering the proton relaxation time (Normann et al., 1995).

Gadolinium (Gd) is a rare earth metal which due to its unpaired electrons is paramagnetic. The free metal is toxic. However, when bound to diethylenetriaminepentaacetic acid (DTPA) or other chelates it is a safe water soluble contrast agent suitable for human use (Westbrook and Kaut, 1993). Gadolinium has similar pharmokinetics as iodinated contrast agents in CT. It also displays a similar volume of distribution and is excreted via the kidneys. It does appear much safer and as yet there have been few reports of serious reactions to Gd DTPA documented.

The paramagnetic properties of gadolinium DTPA when present within a tissue result in a shortened T1 relaxation time and therefore appear as a high signal area on a T1 weighted image (Normann et al., 1995)

Why tumours enhance

Tissue enhancement after an intravenous bolus of Gd DTPA can be displayed as a time signal intensity curve which has four phases, as follows:

- onset of enhancement;
- fastest slope of enhancement;
- the plateau phase;
- the wash-out phase.

A time signal intensity curve indicating these phases is depicted in Figure 18.1. The enhancement rate depends on a number of factors

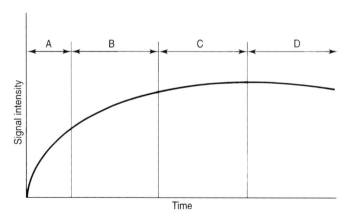

Figure 18.1 A time signal intensity curve demonstrating the four phases of tissue enhancement following a bolus of intravenous gadolinium. A, Onset of enhancement; B, Fastest slope of enhancement; C, Plateau phase; D, Washout phase.

such as: the flow rate, vascular resistance and wall permeability and the composition of the extracellular space. These factors vary with tissue type and therefore affect enhancement curves (Barentsz et al., 1995).

Tumours larger than 2 mm in size secrete angiogenic factors that stimulate the formation of new blood vessels. These regions of hyper-

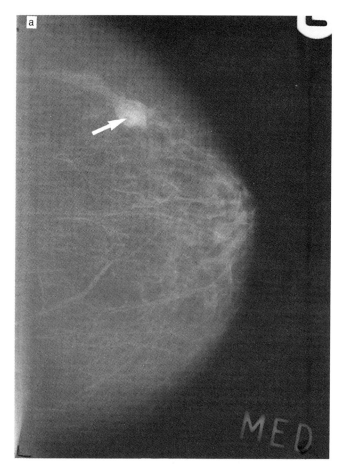

Figure 18.2 (a) The mammographic appearance of a patient who presented with a palpable lesion (arrow) but whose ultrasound was negative. An MRI was performed to identify the nature of the lesion (see Figure 18.2b).

vascularity develop at the periphery of the tumour and create beds for pooling of contrast. Breast carcinomas have been shown to have increased extracellular space and capillary permeability along with expanded interstitial space, all of which result in the accumulation of contrast (Adler and Wahl, 1995). Hypervascularity, however, is also associated with benign conditions such as inflammatory lesions and fibroadenomas, most likely accounting for some of the overlap in enhancement characteristics of different tissues (Adler and Wahl, 1995). This point is demonstrated in Figure 18.2 by a clinical example.

Enhancement patterns

The temporal enhancement varies with tissue type (Figure 18.3). Therefore studies of contrast enhanced breast MRI have focused on the amount, speed and shape of the time signal intensity curve (Adler and

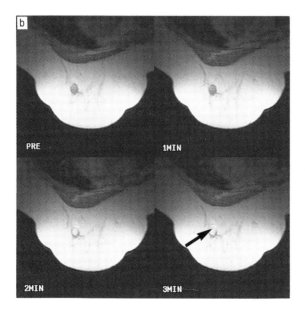

Figure 18.2 (b) The dynamic MRI study demonstrates a well defined rapidly enhancing 1 cm mass in the upper outer quadrant (arrow). The pattern of enhancement is indicative of either a fibroadenoma or a well circumscribed carcinoma. Cytology confirmed the presence of a fibroadenoma. The annotation indicates the time after the bolus of intravenous contrast.

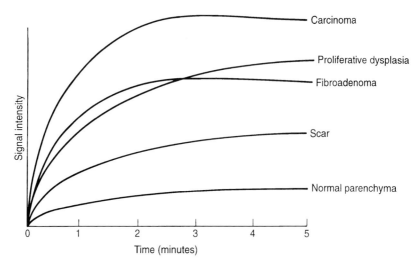

Figure 18.3 Graph demonstrating the variation in temporal enhancement with tissue type.

Wahl, 1995). The pattern of signal enhancement in the first one to two minutes can be used to differentiate between malignant and benign enhancing lesions. Normal breast tissue, non-proliferative dysplasia and old scarring usually do not enhance. Variable amounts of enhancement occur in proliferative dysplasias and inflammatory effects due to fresh postoperative or post radiation changes. All cancers enhance significantly, as do the majority of benign tumours such as fibroadenomas and papillomas and consequently contribute to false positive interpretations (Harms et al., 1994).

The absence or delay in enhancement does exclude malignancy (Heywang-Köbrunner, 1992) as cancers enhance rapidly after injection whereas most benign disorders exhibit delayed or no enhancement as seen in Figure 18.4 (Harms et al., 1994). The difference in appearance between benign and malignant lesions is noticeably reduced several minutes after contrast administration (Kaiser, 1994). Dynamic studies have shown that malignant lesions imaged every 60 seconds for a duration of five minutes following contrast exhibit rapid initial enhancement reaching a plateau after two minutes followed by a rapid decrease in signal intensity which is known as the washout effect. This is in contrast to most benign lesions that demonstrate a more gradual pattern of enhancement (Adler and Wahl, 1995). A dynamic contrast enhanced MRI technique is therefore the most suitable method for differentiating benign and malignant lesions.

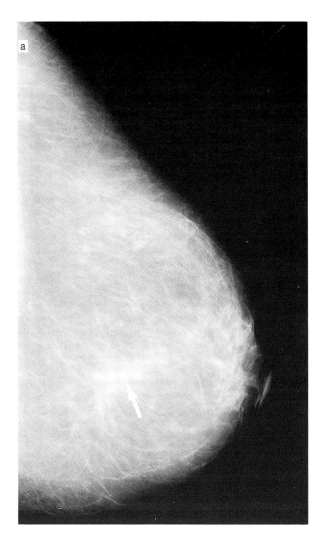

Figure 18.4 (a) A post lumpectomy mammogram image undertaken to ascertain the nature of the lesion (arrow).

TECHNIQUE

Currently there is no single standardized and generally accepted technique for breast MRI with investigators favouring very different protocols. The variance can be attributed to hardware and software

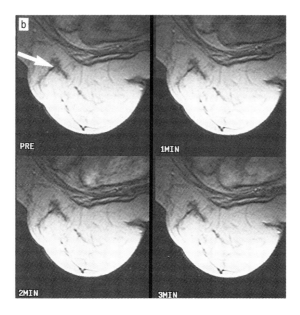

Figure 18.4 (continued)   (b) The pre-contrast image (Pre) of the dynamic study shows an irregular area in the upper outer quadrant (arrow) corresponding with the lesion identified on the mammogram (see Figure 18.4a). There is no enhancement on the post-contrast image (3 min), an appearance indicative of postoperative fibrosis.

capabilities, clinical indications, desired results, personal experience and preferences (Weinreb and Newstead, 1995). The ultimate aim of MRI is to detect small breast lesions and determine whether they are malignant or benign. It is necessary to obtain the imaging information rapidly to ensure that enhancing tumours are discriminated from ductal tissue. The two tissues are easily distinguishable up to five minutes after contrast administration. However, in excess of this time ductal tissue begins to enhance, displaying a similar appearance to cancerous tissue. To maximize the potential of differing temporal enhancement patterns and provide the greatest level of sensitivity, a very high resolution technique imaging the entire breast in just a few seconds is preferable. High resolution combined with short scan times would not necessarily produce the desired image quality as many other factors require consideration. Sensitivity improves with increasing resolution and signal to noise ratio. However, this results in a longer imaging time. Conversely the reduced time required for dynamic scanning reduces the resolution and signal to noise ratio. A suitable balance therefore has to be chosen which often results in a compromise of these factors.

Surface Coils

Coils are used to receive the MRI signal, the smaller the coil and the closer the coil to the body part the better the signal to noise ratio. Utilizing a small diameter coil placed close to the body surface increases the local sensitivity. However, the signal received reduces dramatically with increasing distance from the coil. Breast imaging requires inclusion of the whole breast, the axilla and the chest wall within the sensitive volume of the coil (Hylton and Frankel, 1994). To meet the necessary criteria, developments in design have led to single and bilateral dedicated breast coils arranged in individual wells with padded, anatomically shaped supports. An example of which is pictured in Figure 18.5. These devices allow the patient to lie prone over the coil, enabling comfortable positioning whilst minimizing the effects of respiratory motion.

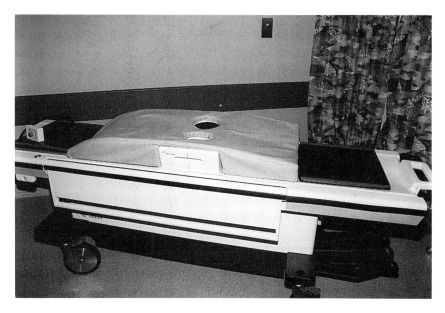

Figure 18.5 A typical dedicated bilateral breast coil on an MRI patient couch. Note the individual wells and cushions for patient comfort.

Patient Positioning

Many patients are anxious therefore reassurance and a full explanation of the procedure is particularly important. The patient lies prone on the imaging couch with the breasts suspended within the breast coil (Figure 18.6). The patient's arms are usually positioned either above the head or at the side ensuring that intravenous access is possible. The intravenous line is inserted before the examination to avoid any patient movement during the study, vital to ensure accurate image registration of the pre- and post-contrast scans (Hylton and Frankel, 1994).

Pulse Sequences

There is general agreement that a T1 weighted sequence must be performed before and after an intravenous injection of gadolinium contrast agent. Spin echo and a variety of gradient echo techniques have been successfully employed (Weinreb and Newstead, 1995). Current gadolinium enhanced studies fall into two categories:

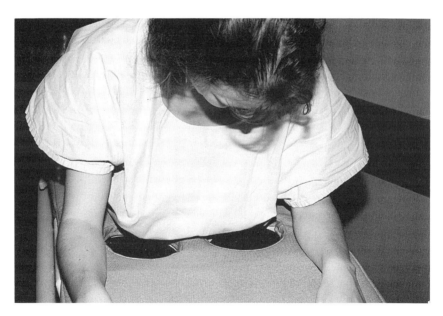

Figure 18.6 (a) Patient positioning for a breast MR study utilizing a dedicated bilateral coil.

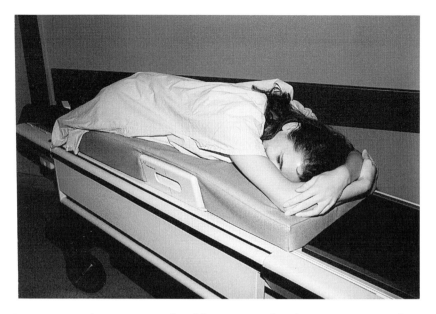

Figure 18.6 (b) Patient comfortably positioned in the prone position for breast imaging.

- limited coverage dynamic studies for characterizing enhancing lesions by their rate of enhancement;
- high resolution 3D large volume studies for detecting small lesions.

T1 weighted gradient echo sequences have replaced T1 weighted spin echo sequences as the technique of choice because of fast scan times necessary for dynamic and three-dimensional (3D) volume acquisitions (Hylton and Frankel, 1994).

Examples of typical parameters for the following described sequences may be found in Table 18.1.

Dynamic contrast enhanced imaging

This involves the acquisition of data for a limited number of slices at short intervals of less than one minute for a total duration of five minutes (Hylton and Frankel, 1994). Gradient echo pulse sequences are capable of acquiring a slice in only 16 seconds. However, this limits the total number of locations to three or four (Figure 18.7a). This potentially reduces MRI's ability to detect multicentric disease or small lesions. Recent technological advances have lead to the development of

Table 18.1 Pulse sequence parameters

Pulse sequence	2D dynamic T1W GE	2D dynamic T1W fast GE	3D volume T1W GE
Scan plane	Axial	Axial	Coronal
Echo time (TE)	5 ms	4.2 ms	14 ms
Repetition time (TR)	35 ms	184 ms	40 ms
Flip angle	35°	60°	50°
Slice thickness	5 mm	5 mm	4 mm
Matrix	192 × 265	128 × 256	128 × 256
Field of view	26 cm	26 cm	26 cm
Number of slices	1–3	16	32
Scan time	16–43 s	49 s	5 min

GE, gradient echo.

fast gradient echo sequences which allow sixteen slice locations to be achieved in under one minute. This enables the whole breast to be imaged (Figure 18.7b) with increased resolution and improved sensitivity whilst maintaining the specificity of a dynamic technique. The data acquired can be used to display time signal intensity curves of specific regions of interest enabling differentiation of benign from malignant lesions. The images can also be displayed as a video loop to enable the differential rates of enhancement to be observed.

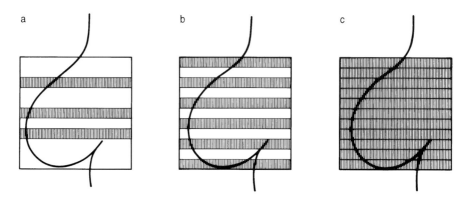

Figure 18.7 Schematic diagram of a sagittal breast demonstrating typical axial slice position and coverage for the following sequences: (a) 2D Gradient echo sequence results in a limited number of slices, therefore they need to be positioned accurately over the areas of interest. (b) 2D Fast gradient echo enables coverage of the whole breast in a very short sequence time. (c) 3D Gradient echo facilitates a high resolution study of the entire breast.

3D volume

The temporal requirements of dynamic scanning restricts the spatial resolution and coverage that are achieved. Improved lesion detection is obtained by utilizing high resolution 3D volume imaging (Figure 18.7c) with approximately 1 mm isotropic voxel resolution. It is recommended that the scan time should be less than five minutes to observe differential enhancement between malignant and ductal tissue (Hylton and Frankel, 1994). A volume acquisition also offers the capability of reformatting the data in a variety of planes allowing improved appreciation of spatial relationships (Weinreb and Newstead, 1995). This technique enables the whole breast to be imaged with moderate specificity and improved sensitivity.

Fat Saturation

A number of studies have found fat saturation to be of value in the identification of breast disease. It has been particularly beneficial when the surrounding normally hyperintense fat on a T1 weighted (T1W) image may obscure the contrast enhanced lesion. It has also been of value in detecting small lesions in fatty breasts where partial volume averaging of fat may obscure lesion enhancement (Hylton and Frankel, 1994). A disadvantage of using a fat saturation technique stems from non-uniformity of the magnetic field which in turn causes imperfect fat suppression potentially obscuring pathological tissue (Weinreb and Newstead, 1995). Enhancing lesions are more evident on T1W incoherent gradient echo images than on those produced by T1W spin echo sequences because of the indeterminate fat signal on gradient echo sequences. Fat saturation is therefore not universally recommended for current breast imaging techniques but tends to be reserved for diagnosing problem cases in patients with fatty breasts (Hickman et al., 1994).

Artefacts

There are numerous artefacts that may degrade MR image quality and breast imaging is no exception. It is important to immobilize the freely hanging breasts as vibratory motion may create movement artefacts which are especially problematic in fast scan techniques. Partial immobilization of the breast without distortion using soft pads is therefore desirable (Hylton and Frankel, 1994). Respiratory and cardiac motion also cause artefacts in the phase encoding direction and are more evident with contrast enhancement. Cardiac gating is usually impractical in breast imaging as it extends the scan times unacceptably. The effects of respiratory motion are minimized as the patient is usually examined in the prone position. To ensure that the phase encoding artefact does not obscure the area of interest the phase axis should be positioned in the right/left direction for axial imaging (Figure 18.8) and in the superior/inferior direction for sagittal imaging. This technique, however, may not be a viable option if the axillary area is being evaluated (Weinreb and Newstead, 1995). To reduce artefact from the heart, pre-saturation bands placed posteriorly on axial scans can be used to saturate the signal from the chest. However care must be taken not to obscure the lateral edges of the breast and axillary tail (Hylton and Frankel, 1994). Phase wrap artefact may also occur if the sensitive volume of the coil extends beyond the field of view selected. This is a particular problem when imaging in the sagittal plane and the phase and frequency directions have been swapped. Increasing the field of view size or oversampling in the phase direction, frequently termed no phase wrap, can overcome this but the penalty is reduced resolution or increased scan time (Hylton and Frankel, 1994).

APPLICATIONS

It has been suggested that magnetic resonance imaging may have a role in the detection and management of breast disease. However, because there has been little standardization in the techniques used there is still uncertainty of its place in clinical practice.

A study by the Breast Cancer Demonstration Project showed that approximately 9% of palpable carcinomas were not seen on mammography. In another MR study of 18 patients with breast cancer, the planned therapy or surgical approach was changed in half of the patients. This alteration of treatment directly resulted from new information provided by MRI on size, multifocality, multicentricity or chest wall invasion. It is therefore thought that MRI may provide the capability of filling the current gap between the information available from conventional imaging and clinical requirements (Harms et al., 1992). This is particularly important as breast conservation surgery is placing increasing demands on imaging techniques to accurately stage the disease (Harms et al., 1992).

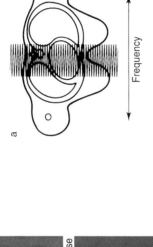
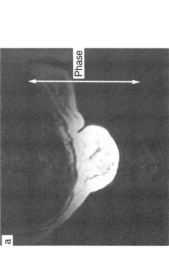

Figure 18.8 The effects on image quality of altering the phase and frequency direction are demonstrated. (a) Phase is anterior/posterior resulting in the artefact obscuring the area of interest.

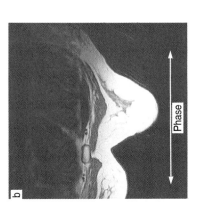
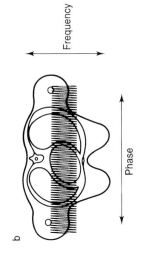

Figure 18.8 (continued) (b) Phase is in the right/left direction resulting in improved image quality of the breast area; however, the artefact may now obscure the axilla.

Assessment of Multifocal and Multicentric Disease

It has been suggested that MRI has a role in determining and planning treatment for those women who desire breast preservation, as determining the extent of the disease is particularly problematic using conventional imaging techniques. Multiple breast cancers reportedly occur in 9–75% of patients with non-palpable breast cancer detected on mammography (Harms et al., 1992). Many of these, however, are histologically microscopic and non-invasive and following radiotherapy will probably not develop into a clinical problem.

It is also important to distinguish between multifocal and multicentric disease since this affects the patient's suitability for breast conservation therapy. Multifocal disease is defined as multiple cancers derived from the same primary tumour in a single quadrant, whereas multicentric disease involves cancers in separate quadrants of the breast, these being unsuitable for breast conservation therapy.

MRI could be used to classify patients with breast cancer into three groups (Weinreb and Newstead, 1995), as follows:

- unifocal disease that requires only lumpectomy without radiation;
- minimal multifocal disease where quadrantectomy with radiation would be sufficient;
- extensive multifocal or multicentric disease prone to recurrence and where mastectomy would be recommended.

Radiation therapy is used to supplement the minimal surgery in breast conservation by aiming to eradicate residual microscopic foci. This staging enables the cosmetic appearance of the breast to be maintained and is based on the assumption that the remainder of the breast contains only microscopic focus. The sensitivity of imaging is therefore vital in this group as even after the excision of the clinically apparent tumour, residual subclinical cancerous tissue in the breast can be considerable (Harms et al., 1992).

Dynamic contrast enhanced MRI of the whole breast is particularly useful in improving the sensitivity for the detection and staging of multicentric lesions often missed by conventional techniques (Harms et al., 1992). Recent studies confirm that MRI may perform better than mammography; for example, in a study of 47 patients, agreement with mammography was only reached in 64% of patients. MRI detected twice the number of lesions as mammography and of those lesions that were not identified 62% were malignant. The majority of additional lesions detected by MRI represented multifocal disease in patients with only a solitary focus on mammography (Harms et al., 1992). Figure 18.9 depicts a clinical example of multifocal disease on MRI where the mammogram demonstrated a single opacity.

Mammographically Dense Breast

The majority of tumours that are missed on screening mammograms are likely to occur in dense rather than fatty breasts. Women with mammographically dense breasts make up 25% of the population and present a challenge in the detection of early stage carcinoma. The difficulty in imaging the radiologically dense breast and determining tissue characteristics stems from a number of factors (Jackson et al., 1993).

These are as follows:

- breast lesions have x-ray attenuation properties similar to those of dense glandular and fibrous tissues, reducing the probability that a lesion will be clearly visible in the dense breast (Jackson et al., 1993);
- the dense breast produces more 'scatter', reducing image contrast;
- there are greater tissue inhomogeneities and a wider attenuation range making it difficult to optimally expose all areas of the breast (Jackson et al., 1993).

MRI is considered to show significant promise for imaging the dense breast as the radiological problems do not apply to dynamic contrast enhanced MRI (Jackson et al., 1993).

Recurrence versus Scar

MRI of the breast has been found to be useful in identifying tumour recurrence in women treated with lumpectomy and radiotherapy. Some studies have shown that whilst recurrent tumours enhance, postoperative scars older than six months do not demonstrate the same enhancement (Adler and Wahl, 1995). MRI accurately identifies scar tissue (Figures 18.10 and 18.11) at the site of a palpable mass or mammographic lesion (Hickman et al., 1994) although in some cases there may be non-specific enhancement up to 18 months after surgery or radiotherapy (Weinreb and Newstead, 1995). These findings are important as the alternative of aspiration cytology requires expert assess-

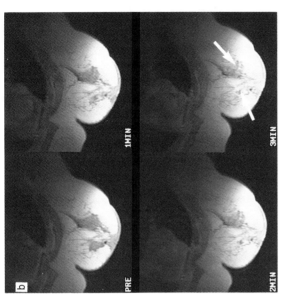

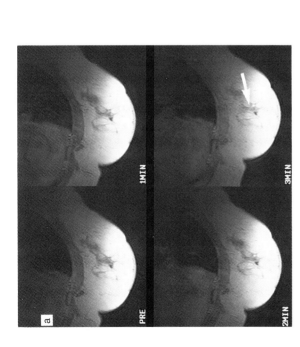

Figure 18.9 (a) and (b) Images of two slice locations from an axial dynamic MR study one year after a partial mastectomy, adjuvant chemo and radiotherapy. The routine mammogram (not shown) depicted a poorly defined opacity which had the appearance of either scar tissue or tumour recurrence. The pre-contrast MR study demonstrates a lobulated mass in the upper inner quadrant and on the post-contrast image the mass shows rapid pronounced enhancement (arrows), indicative of multifocal malignancy.

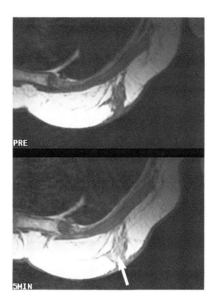

Figure 18.10 Images from an axial dynamic study undertaken to clarify the nature of a small lump at the medial end of the scar from a partial mastectomy performed two years previously. The pre-contrast image demonstrates thickening of the scar and an irregular heterogeneous mass. The scar shows marked enhancement (arrow) on the post-contrast images, exhibiting the typical features of local recurrence with deep extension.

ment and may still be equivocal in cases that have had external radiotherapy. Biopsies also have other disadvantages, for example they are invasive, which may result in bleeding and stimulate tumour growth, and they may compound the distortion of the breast as well as causing anxiety (Hickman et al., 1994).

Silicone Implants

Breasts with silicone implants pose problems for mammographic examination due to the high radiographic density of silicone that may obscure up to 80% of breast tissue. The prosthesis, however, does not impair cancer visualization on MRI images (Harms et al., 1994). MRI is also used to accurately identify silicone leaks that may be mistaken for other masses (Harms et al., 1992). This application has been most

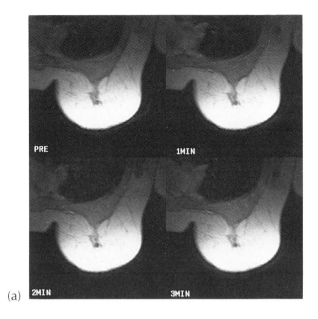

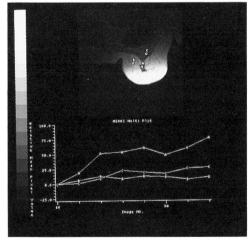

Figure 18.11 Images from an axial dynamic study along with a time signal intensity curve undertaken to identify a firm swelling over a scar from wide local excision of carcinoma followed by radiotherapy one year previously. (a) There is little evidence of enhancement on the post gadolinium image which is confirmed on (b) the time signal intensity curve. The small amount of lesion enhancement, annotated as 'A' on the graph depicts a slow minimal rate of enhancement indicative of fibrosis.

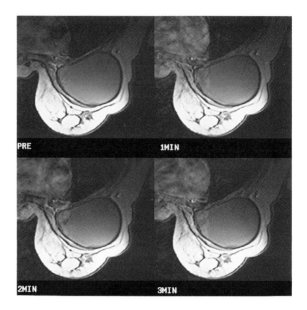

Figure 18.12 Images from an axial dynamic study demonstrating the MR appearance of a silicone breast prosthesis extending into the pectoralis major muscle. There is no enhancement visible within the normal breast tissue excluding malignancy.

widespread in the USA in pursuit of medicolegal claims. The MRI appearance of a silicone implant is demonstrated in Figure 18.12.

Occult Cancer

MRI of the breast may be useful in women with occult cancer (Figure 18.13). This group often presents with axillary adenopathy from adenocarcinoma but do not demonstrate mammographic, sonographic or palpable evidence of a primary carcinoma in the breast or elsewhere.

Mastectomy has usually been the treatment of choice despite the fact that a primary tumour in the breast will be found in only two out of every three patients. If the occult cancer could be localized with MRI, the site ambiguity of the primary tumour would be eliminated enabling a limited surgical procedure combined with adjuvant therapy (Weinreb and Newstead, 1995).

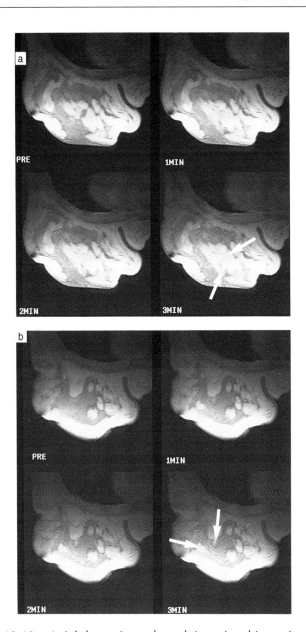

Figure 18.13 Axial dynamic study and time signal intensity curve performed to investigate a clinical presentation of occult carcinoma with bulky nodes in the left axilla and a negative mammogram. The objective of the MRI was to locate the primary. The images at two slice locations (a) and (b) of the dynamic study demonstrate several foci of abnormal enhancement (arrows), features typical of malignancy.

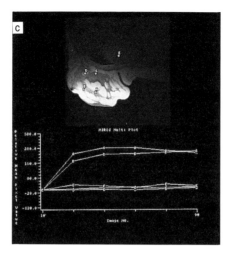

Figure 18.13 (continued)   (c) The time signal intensity curve of another slice depicts two lesions, labelled A and B on the graph possessing an enhancement pattern typical of malignancy. These findings were confirmed at mastectomy.

Difficult Cases

MRI is also useful in providing the information required for effective lumpectomy by accurately defining the extent and margins of a mass (Harms et al., 1992).

The superior sensitivity of MRI makes it an ideal technique for patients who are strongly suspected of having breast cancer, but for whom conventional imaging studies have failed to demonstrate an abnormality (Harms et al., 1992).

Other pathologies that are well demonstrated on MRI include haematoma (Figure 18.14) following a biopsy and surgery.

Limitations

Despite the proven advantages of MRI there are also a number of limitations that require emphasizing:

- MRI cannot fully replace biopsy due to its inability to discriminate consistently between benign and malignant lesions (Harms et al., 1992).

MRI of the Breast 223

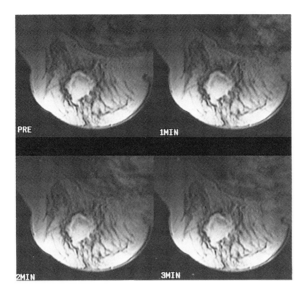

Figure 18.14 MR images demonstrating the appearance of a haematoma. An irregular multilocular mass can be seen which displays a heterogeneous high signal on both T1 and T2 weighted images accompanied by a thick low signal rind and fluid level. No enhancement is depicted following gadolinium.

- False positives
  - some benign masses enhance with contrast and may be difficult to distinguish from malignancy;
  - other benign conditions that enhance potentially misleading diagnosis include normal glandular tissue depending on the phase of the menstrual cycle, and post biopsy oedema (Harms and Flamig, 1992).
- False negatives
  - false negative MR findings after gadolinium enhancement have been attributed to technical factors such as artefacts, slice gaps or misinterpretation;
  - insufficient doses of contrast agent and instances when carcinoma size is smaller than the slice thickness, have also caused false negatives (Adler and Wahl, 1995).

Cost

Conventional breast imaging, mammography and ultrasound are all considered low cost clinical investigations.

MRI, however, is inherently more expensive and is highly unlikely to be competitive with mammography as the gadolinium alone would cost more than two mammogram examinations (Harms and Flamig, 1993). The sole use of MRI as the investigation of choice, despite improved sensitivity, is difficult to justify.

A fine needle biopsy is considerably less expensive than MRI and provides a definite cytological diagnosis (Harms et al., 1992). MRI therefore must provide some important information that cannot be obtained with widely available, less expensive technology if it is to become a practical option in the clinical detection of breast cancer (Harms and Flamig, 1993).

MRI GUIDED BIOPSY

Mammography guided stereotactic biopsy is a well established technique for accurately determining tissue pathology. Ideally a suspicious tumour identified on an MRI study is biopsied and localized for surgical intervention. Surgical treatment, however, cannot be implemented with absolute confidence on the basis of MRI findings as the technique offers only limited prospects of discriminating between the various pathologies (Kerslake et al., 1995). Since lesions identified by MRI alone cannot be localized by conventional methods there is currently no method for clinically addressing this problem (Harms et al., 1992). Free hand placement of a needle into the area of concern is very difficult because contrast enhancement diminishes with time.

A method for stereotactic localization and biopsy is required if MRI is to be of true clinical value.

Two methods are currently being investigated:

- The use of a breast localization coil. The patient lies prone and the breast is compressed between plates containing a total of one thousand 18 gauge holes. The breast is imaged and the patient is then withdrawn from the bore of the magnet. The lesion is localized and utilizing the coordinates obtained from the images the needle is guided through a preformed hole in the biopsy plate into the lesion. The patient may then receive repeat imaging for confirmation of

needle positioning prior to the biopsy (Weinreb and Newstead, 1995).
- The use of a mechanical stereotactic external needle guide. This method utilizes the MR equipment's x, y and z coordinates to position the needle accurately within the breast (Greenstein et al., 1994).

The techniques described both result in a substantial artefact caused by the non-ferromagnetic biopsy needle, making it difficult to confirm the position of the needle tip within a small lesion.

CONCLUSION

It is widely considered that MRI has the potential to bridge the gap between the clinical information required for appropriate treatment and that currently available from other imaging methods (Harms et al., 1992). The increased role of breast conservation surgery, for example, is demanding more accurate staging and definition providing an area where MRI may establish its role.

There is no doubt that the extent of tumours is more readily appreciated on MRI compared to conventional assessment and is particularly useful in patients with dense breasts (Kerslake et al., 1995). Contrast enhanced MRI can improve the specificity of breast imaging by its ability to distinguish between unenhancing benign lesions such as scar or fat necrosis and enhancing tumour. There may, however, be non-specific enhancement of tissue after surgery and radiotherapy. The distinction between benign enhancing lesions and malignant tissue, however, is more difficult to ascertain so it is unlikely that MRI will be able to replace pathological examination of tissue biopsies (Harms et al., 1992).

MRI is inherently expensive but if it is able to minimize inappropriate surgery it could become a cost-effective technique and reduce morbidity (Kerslake et al., 1995).

THE FUTURE OF MRI IN BREAST IMAGING

It is difficult to predict the future role of MRI in breast diagnosis as many of its applications, despite showing promise, are still in the investigational stage. The shortcomings of mammography are widely acknowledged but there is no consensus regarding the clinical role of MRI. There is widespread scepticism of results which is mainly due to poor standardization of techniques and equipment resulting in a lack of comparability between studies. The credibility of MR would also improve if standards of quality control were developed for all aspects of the technique to include image interpretation.

Unfortunately despite the interest in breast MR, large controlled trials with pathological, mammographic and MR correlation have not been undertaken. Evidence from such a trial may establish the role of breast MR but it is probable that the two remaining hurdles will be specificity and cost.

There are, however, promising aspects that may improve its acceptance. Technological advances may result in shorter examination times whilst improving accuracy. Cheaper, dedicated MR systems may become available which in conjunction with high throughput will reduce examination cost and increase availability.

It seems most likely that MRI will not replace mammography but be undertaken in conjunction with it to answer the more problematic clinical questions. MRI as a supplementary tool will help to increase both specificity and sensitivity.

Acknowledgements

The authors wish to thank Dr N. R. Moore for advice and clinical cases and Sarah Mitchell for her secretarial assistance. General thanks are due to Dr M. J. Warren and the staff at the MRI Centre, John Radcliffe Hospital, Oxford.

References

Adler, D.D., Wahl, R.L. (1995) New methods for imaging the breast: techniques, findings, and potential. *American Journal of Roentgenology* **164:** 19–30.

Barentsz, J.O., Boetes, C., Verstraete, K.L. et al. (1995) Dynamic gadolinium-enhanced MR imaging of the body. *Clinical MRI* **5(3):** 88–93.

Bieze, J., Ward, P. (1995) Breast MR needs data to soothe the growing pains. *Diagnostic Imaging Europe* **33:** 37, 51.

Cancer Research Campaign (1991) Factsheet 6. 1–3.

Cosgrove, D.O., Bamber, J.C., Davey, et al. (1990) Color Doppler signals from breast tumours: work in progress. *Radiology* **176:** 175–180.

Dixon, J.M. (1993) Breast Conservation surgery. In Taylor, I., Johnson, C.D. (eds) *Recent Advances In Surgery – 16*, 1st edn, pp. 43–61. Edinburgh: Churchill Livingstone.

Greenstein, Orel S., Schnall, M.D., Newman, R.W. et al. (1994) MR imaging-guided localization and biopsy of breast lesions: initial experience. *Radiology* **193**: 97–102.

Harms, S.E., Flamig, D.P. (1993) MR imaging of the breast. *Journal of Magnetic Resonance Imaging*, **3**: 277–283.

Harms, S.E., Flamig, D.P., Heslet, K.L. et al. (1992) Magnetic resonance imaging of the breast. *Magnetic Resonance Quarterly* **8(3)**: 139–155.

Harms, S.E., Flamig, D.P., Evans, W.P. et al. (1994) MR imaging of the breast: current status and future potential. *American Journal of Roentgenology* **163**: 1039–1047.

Heywang-Köbrunner, S.H. (1990) Conventional breast diagnostics. In Heywang-Köbrunner, S.H. (ed) *Contrast-Enhanced MRI of the Breast,* 1st edn, pp. 12–15. Munich: Karger Basel.

Heywang-Köbrunner, S.H. (1992) Non mammographic breast imaging techniques. *Current Options in Radiology* **4(V)**: 146–154.

Hickman, P.F., Moore, N.R., Shepstone, B.J. (1994) The indeterminate breast mass. Assessment using contrast enhanced MRI. *British Journal of Radiology* **67**: 14–20.

Homer, M.J. (1985) Breast imaging: Pitfalls, controversies and some practical thoughts. *Radiological Clinics of North America* **23/3**: 459–472.

Hylton, N.M., Frankel, S.D., (1994) Imaging Techniques for Breast MR Imaging. In Davies, P.L. (ed) *Magnetic Resonance Imaging MRI Clinics of North America, Breast Imaging Vol. 2. No. 4*, pp. 511–524. Philadelphia: W.B. Saunders.

Jackson, V.P., Hendrick, R.E., Feig, S.A. et al. (1993) Imaging of the radiographically dense breast. *Radiology* **188**: 297–301.

Kaiser, W.A. (1994) False-Positive results in Dynamic MR Mammography. In Davies, P.L. (ed) *Magnetic Resonance Imaging MRI Clinics of North America, Breast Imaging Vol. 2. No. 4*, pp. 539–555. Philadelphia: W.B. Saunders.

Kerslake, R.W., Carleton, P.J., Fox, J.N. et al. (1995) Dynamic gradient-echo and fat-suppressed spin-echo contrast enhanced MRI of the breast. *Clinical Radiology* **50**: 440–454.

Mintun, M.A., Welch, M.J., Siegel, B.A. et al. (1988) Breast cancer; PET imaging of estrogen receptors. *Radiology* **169**: 45–48.

Normann, P.T., Hustvedt, S.O., Storflor, H. et al. (1995) Preclinical safety and pharmacokinetic profile of gadodiamide injection. *Clinical MRI* **5(3)**: 95–101.

Sluyser, M., Hoefnagel, C.A. (1988) Breast carcinomas detected by thallium-201 scintigraphy. *Cancer Letters* **40**: 161–168.

Wald, N.J., Murphy, P., Major, P. et al. (1995) UKCCCR multicentre randomised controlled trial of one and two view mammography in breast cancer screening. *British Medical Journal* **311**: 1189–1193.

Warren, S.L. (1930) Roentgenologic study of the breast. *American Journal of Roentgenology* **24**: 113.

Weinreb, J.C., Newstead, G. (1995) MR imaging of the breast. *Radiology* **196**: 593–610.

Westbrook, C., Kaut, C. (1993) Contrast Agents in MRI. In Westbrook, C., Kaut, C. (eds) *MRI in Practice,* 1st edn, pp. 241–260. Oxford: Blackwell Scientific Publications.

19

Coronary Artery Imaging by Magnetic Resonance

Dudley Pennell

INTRODUCTION

Coronary artery disease is a leading cause of death and disability in the Western world, and in the UK in 1985 was responsible alone for 163 104 (or 33%) of all deaths (Office of Health Economics, 1987). The management of coronary artery disease depends on a subjective assessment of patient symptoms and trials of medical therapy, an objective assessment of prognosis, and an objective appraisal of the likelihood of success of any physical intervention such as angioplasty or bypass surgery. Coronary artery imaging is important for both the assessment of prognosis and the possibilities for intervention. Although very significant information regarding prognosis can now be gained from myocardial perfusion imaging (Pennell and Prvulovich, 1995), which is similar or superior to that of the coronary anatomy alone (Iskandrian et al., 1993), ultimately intervention still requires direct imaging of the location of stenosis, and the suitability of the distal arterial lumen for bypass. The only technique available for this until very recently was conventional x-ray coronary angiography, which involves intra-arterial catheterization, significant x-ray and contrast medium exposure, hospital admission, and a small but real risk of significant morbidity and mortality. Now however, a new technique using magnetic resonance (MR) imaging is under development which in time may have a significant impact on the routine practice of the conventional x-ray technique.

Progress using magnetic resonance angiography of the coronary vessels has been hindered by a combination of formidable problems. These include their small calibre, cardiac motion, respiratory motion, tortuosity, and proximity to other tissues of high proton density. These problems are now being addressed by the use of fast acquisition techniques, fat suppression and breath-holding.

MR IMAGING TECHNIQUES

Early coronary imaging, using gated spin echo or gradient echo sequences with a four minute acquisition time showed the proximal coronary arteries in some patients (Underwood, 1991) but a robust technique which reduced blurring and artefact could not be developed (Paulin et al., 1987). The recent development which has made all the difference has been the introduction of a segmented k-space technique (Manning et al., 1993a). The acquisition of k-space is performed in blocks with this technique, such that imaging time may be cut by a factor of 8 or greater. This has reduced the imaging time for single images to under 20 seconds during which it is possible to breath-hold and eliminate respiratory movement artefact. Image acquisition takes approximately 100 ms, and when this is gated to mid-diastole there is also considerable reduction in cardiac motion. Finally the signal from fat is eliminated with a suppression pulse, and this allows the luminal water signal to be identified from the high surrounding signal of epicardial adipose. This technique has also been used by Pennell et al. (1993), Duerinckx and Urman (1994) and van Rossum et al. (1995). Saturation of myocardium has also been tried using magnetization transfer contrast to improve the signal contrast between myocardium and coronary luminal signal (Li et al., 1993).

Other ultrafast imaging techniques have also been evaluated but are not widely available. These include echo planar imaging (Stehling et al., 1987), spiral imaging (Meyer et al., 1992) subtraction techniques (Wang et al., 1991, Paschal et al., 1993), multiple thin slice 2D acquisitions with postprocessing (Cho et al., 1991), and 3D acquisitions (Li et al., 1993).

To follow the course of the coronary arteries multiple contiguous slices are acquired, each in a separate breath-hold. This can be in transaxial or longitudinal planes. Problems with the reproducibility of the depth of breath-hold can lead to apparent discontinuities which could be misinterpreted as signal loss at a stenosis, and respiratory monitoring is being evaluated to combat this. The technique most currently favoured is that of MR diaphragm imaging prior to acquisition, such that accurate positioning is maintained, with a technique known as navigator echoes (Liu et al., 1993). The other value to such monitoring is that ultimately it should allow the elimination of breath-holding, which presents few problems to normal subjects, but is less robust in patients. An alternative approach is a breath-hold scheme requiring more frequent but shorter breath-hold periods, such as respiratory cycling with a prolonged expiration phase (Doyle et al., 1993). This novel approach introduces a progressive distortion in the position of the coronary arteries in the 3D-data set, but registration between individual slices is excellent and acquisition is relatively fast and standardized.

Acquisition of high resolution images allows measurements of the vessel lumen, but lowers the signal to noise level. Although signal can be maintained using surface coils, this introduces difficulties with the posterior vessels which are not well seen. At present, typical in-plane resolution is 0.8×1.6 mm.

VALIDATION OF MRI MEASUREMENTS AND COMPARISON WITH X-RAY CORONARY ANGIOGRAPHY

There remain few direct comparisons between MRI and x-ray coronary angiography in normals (Manning et al., 1993a; Pennell et al., 1993) and in coronary stenosis (Manning, 1993b; Pennell et al., 1994; Duerinckx and Urman, 1994; van Rossum et al., 1995). Despite the preliminary nature of the reports, the MR identification and measurements of coronary arteries show good correlation to those from conventional x-ray angiography. It was shown (Pennell et al., 1994) that the proximal arteries were clearly identified in nearly all patients with the exception of the left circumflex artery, which is difficult to demonstrate because of its posterior location (left main stem (LMS) 95%, left anterior descending artery (LAD) 91%, right coronary artery (RCA) 95%, and left circumflex artery (LCx) 76%). These are very similar results to those obtained by Manning et al. (1993a). Measurements of the proximal arterial diameters and comparison with reference values from post-mortem studies showed no significant differences (Dodge et al., 1992) (Table 19.1).

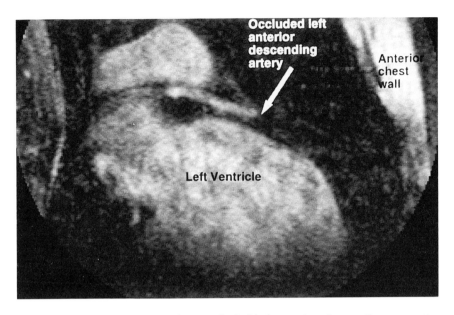

Figure 19.1 MR imaging of an occluded left anterior descending artery in an oblique sagittal plane, similar to a right anterior oblique projection used in conventional x-ray coronary angiography (reproduced with permission from *British Heart Journal* 1993; **70**: 315-326).

Table 19.1 Coronary arteries diameter measured by MRI compared with reference (LMS – left main stem, LAD – left anterior descending artery, LCx – left circumflex artery, RCA – right coronary artery)

Mean diameter (mm) (SD)	LMS	LAD	RCA	LCx
Reference diameters	4.5 (0.5)	3.6 (0.5)	3.9 (0.6)	3.4 (0.5)
Manning et al. (1993b)	4.8	3.6	3.7	3.5
Pennell et al. (1993)	4.8 (0.8)	3.7 (0.5)	3.9 (0.9)	2.9 (0.6)
Duerinckx and Urman (1994)	5.2 (0.9)	4.0 (0.8)	4.7 (0.9)	3.4 (0.9)

Comparison of the two techniques in measurement of the diameter of the proximal coronary arteries can also be made using the width of the arterial catheter for scale on the x-ray film. In the author's experience (Pennell et al., 1993), the mean proximal arterial diameter measured by MRI was 3.9 (1.1) mm compared with 3.7 (1.0) by x-ray angiography (p = NS). Regression analysis has also shown good results (r = 0.86, p < 0.001) (Manning et al., 1993a). The length of the coronary arteries

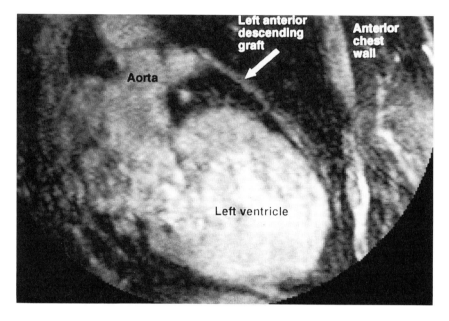

Figure 19.2 MR imaging of an occluded left anterior descending artery and bypass graft in an oblique sagittal plane (reproduced with permission from *British Heart Journal* 1993; **70**: 315–326).

visualized by MRI has been very good considering their known size and tortuosity (Table 19.2). Again, visualization of the LCx suffered from its posterior location.

In the detection of coronary artery disease, Manning et al. (1993b) studied 39 patients and used the criteria of marked attenuation of luminal diameter or signal void in the determination of abnormality of MR coronary angiography. Blinded comparison with x-ray angiography showed a 90% sensitivity and 92% specificity for the detection of moderate and severe coronary stenosis. In a highly selected population, 85% of coronary stenoses were detected (Pennell et al., 1994)

Table 19.2 Lengths of coronary arteries visualized by MRI (abbreviations as in Table 19.1)

Mean length (mm)	LAD	RCA	LCx
Manning et al. (1993b) (range)	44 (28–93)	58 (24–122)	25 (9–42)
Pennell et al. (1993) (range)	47 (14–94)	54 (11–118)	26 (5–63)
Duerinckx and Urman (1994)	53	65	19

but Duerinckx and Urman (1994) detected only 63%, and van Rossum et al. (1995) 41%. Specificity values have also varied. These differences in results are explicable in terms of the patient populations studied (for example, selected versus consecutive patients), and the varying hardware and acquisition techniques. However, the variability shows that magnetic resonance coronary angiography requires considerable further development before widespread clinical application.

ASSESSMENT OF CORONARY STENOSIS

The author has validated the MR technique for assessing coronary stenosis by comparing MR and X-ray for the distance to the stenosis from the arterial origin, and the length and severity of the stenosis (Pennell et al., 1994). The distance measurements showed very high agreement, with the distance to 83% of stenoses being concordant to within 5 mm. There was a significant agreement as well between the degree of MR signal loss and the severity of the stenosis by X-ray, but scatter was higher in these results. There was also a reasonable agreement for the length of the stenosis, but this worsened as stenosis severity increased, suggesting that turbulence was causing overestimation.

The current difficulties and the need for further development has not stopped the use of magnetic resonance coronary angiography in

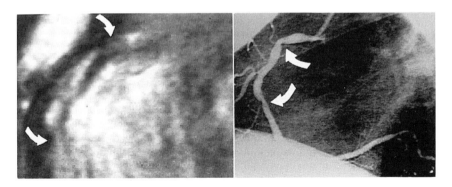

Figure 19.3 MR imaging (left) in an oblique sagittal plane of right coronary artery which has two stenoses, with abnormal signal loss shown by the curved arrows. The corresponding x-ray angiogram (right) is shown for comparison. The stenoses are indicated by arrows.

selected clinical circumstances. This includes the identification of anomalous coronary arteries, the demonstration of coronary patency in patients with occluded native vessels and severe bypass graft disease to assist in the decision to proceed to further bypass grafting, and after infarction for the non-invasive determination of the patency of the infarct related artery following thrombolysis (Hundley et al., 1995). Coronary occlusion and imaging of vein grafts by breath-hold techniques is also straightforward.

IMPROVEMENT OF STENOSIS ASSESSMENT

Although the degree of signal loss at the site of stenosis is dependent on the stenosis severity and the imaging parameters, current experience would not suggest that stenosis can be confidently assessed by current MR techniques. An alternative may be provided by measuring changes in coronary flow velocity between the proximal normal vessel and the stenosis site, because as the area of an artery decreases at a stenosis, the velocity increases in direct proportion. This has been shown using Doppler guide wires in vivo. Early results are promising, but considerable validation work remains (Keegan et al., 1994). In addition, flow quantification by MR including estimation of the coronary flow reserve is on the threshold of becoming a clinical reality, and this would further allow a non-invasive assessment of stenosis severity (Edelman et al., 1993; Clarke et al., 1995).

FUTURE DIRECTIONS OF MR CORONARY ANGIOGRAPHY

The features of a clinically useful MR examination might include: improved resolution, reasonable cost, a standardized imaging protocol completed in under 30 minutes, accurate detection and localization of stenosis and a quantitative evaluation of stenosis severity. Advanced features would include an assessment of plaque lipid content, real time imaging, open access (patient friendly) magnets, and the capability for intervention. The determination of velocity at the stenosis site and the coronary flow reserve from velocity mapping may also prove clinically useful in assessing stenosis severity. Ultimately, however, such an approach might be supplanted by quantitative myocardial perfusion imaging with MRI, which is also under development (Wilke et al., 1994).

Acknowledgements

This work was supported in part by grants from the British Heart Foundation and CORDA (The Coronary Artery Disease Research Association).

References

Cho, Z.H., Mun, C.W., Friedenberg, R.M. (1991) NMR angiography of coronary vessels with 2-D planar image scanning. *Magnetic Resonance Medicine* **20:** 134–143.

Clarke, G.D., Eckels, R., Chaney, et al. (1995) Measurement of absolute epicardial coronary artery flow and flow reserve with breath-hold cine phase-contrast magnetic resonance imaging. *Circulation* **91:** 2627–2634.

Dodge, J.T., Brown, B.G., Bolson, E.L., Dodge, H.T. (1992) Lumen diameter of normal coronary arteries. Influence of age, sex, anatomic variation, and left ventricular hypertrophy or dilatation. *Circulation* **86:** 232–246.

Doyle, M., Scheidegger, M.B., de Graaf, R.G. et al. (1993) Coronary artery imaging in multiple 1-sec breath holds. *Magnetic Resonance Medicine* **11:** 3–6.

Duerinckx, A.J., Urman, M.K. (1994) Two-dimensional coronary MR angiography: analysis of initial clinical results. *Radiology* **193:** 731–738.

Edelman, R.R., Manning, W.J., Gervino E., Li, W. (1993) Flow velocity quantification in human coronary arteries with fast breath-hold MR angiography. *Journal of Magnetic Resonance Medicine* **3:** 699–703.

Hundley, W.G., Clarke, G.D., Lansau, C. et al. (1995) Noninvasive determination of infarct artery patency by cine magnetic resonance angiography. *Circulation* **91:** 1347–1353.

Iskandrian, A.S., Chae, S.C., Heo, J. et al. (1993) Independent and incremental prognostic value of exercise single photon computed emission tomographic (SPECT) thallium imaging in coronary artery disease. *Journal of the American College of Cardiologists* **22:** 665–670.

Keegan, J., Firmin, D., Gatehouse, P., Longmore, D. (1994) The application of breath-hold phase velocity mapping techniques to the measurement of coronary artery blood flow velocity: phantom data and initial in vivo results. *Magnetic Resonance Medicine* **31:** 526–536.

Li, D., Paschal, C.B., Haacke, E.M., Adler, L.P. (1993) Coronary arteries: three dimensional MR imaging with fat saturation and magnetization transfer contrast. *Radiology* **187:** 401–406.

Liu, Y.L., Riederer, S.J., Rossman, P.J. et al. (1993) A monitoring, feedback, and triggering system for reproducible breath-hold MR imaging. *Magnetic Resonance Medicine* **30:** 507–511.

Manning, W.J., Li, W., Boyle, N.G., Edelman, R.R. (1993a) Fat-suppressed breath-hold magnetic resonance coronary angiography. *Circulation* **87:** 94–104.

Manning, W.J., Li, W., Edelman, R.R. (1993b) A preliminary report comparing magnetic resonance coronary angiography with conventional angiography. *New England Journal of Medicine* **328:** 828–832.

Meyer, C.H., Hu, B.S., Nishimura, D.G., Macovski, A. (1992) Fast spiral coronary artery imaging. *Magnetic Resonance Medicine* **28:** 202–213.

Office of Health Economics (1987) *Coronary Heart Disease: The Need for Action*. London: Office of Health Economics.

Paschal, C.B., Haacke, E.M., Adler, L.P. (1993) Three dimensional MR imaging of the coronary arteries: Preliminary clinical experience. *Journal of Magnetic Resonance Imaging* **3:** 491–500.

Paulin, S., von Schulthess, G.K., Fossel, E., Krayenbuehl, H. (1987) MR imaging of the aortic root and proximal coronary arteries. *American Journal of Roentgenology* **148:** 665–670.

Pennell, D.J., Prvulovich, E. (1995) Prognosis in coronary artery disease. In *Nuclear Cardiology*, pp. 95–117. London: British Nuclear Medicine Society.

Pennell, D.J., Keegan, J., Firmin, D.N. et al. (1993) Magnetic resonance imaging of coronary arteries; technique and preliminary results. *British Heart Journal* **70:** 315–326.

Pennell, D.J., Bogren, H.G., Keegan, J. et al. (1994) Detection, localisation and assessment of coronary artery stenosis by magnetic resonance imaging. *Proceedings of the Society of Magnetic Resonance* 369.

Stehling, M., Howseman, A., Chapman, B. et al. (1987) Real-time NMR imaging of coronary vessels. *Lancet* **ii:** 964–965.

Underwood, S.R. (1991) Imaging of acquired heart disease. In Underwood, S.R., Firmin, D.N. (eds) *Magnetic Resonance of the Cardiovascular System*, pp. 41–67. London: Blackwell.

van Rossum, Post, J.C., Hofman, B.M. et al. (1995) Current limitations of two-dimensional breath-hold magnetic resonance coronary angiography (abstract). *Journal of the American College of Cardiologists* 134A.

Wang, S.J., Hu, B.S., Macovski, A., Nishimura, D.G. (1991) Coronary angiography using fast selective inversion recovery. *Magnetic Resonance Medicine* **16:** 417–423.

Wilke, N., Jerosch-Herold, M., Stillman, A.E. et al. (1994) Concepts of myocardial perfusion in magnetic resonance imaging. *Magnetic Resonance Quarterly* **10:** 249–286.

20
The Role of Ultrasound in the Diagnosis and Management of Scoliosis

Alanah S. Kirby

INTRODUCTION

Scoliosis

Scoliosis is a lateral curvature of the spine which can cause asymmetry of shoulders and hips. The lateral curvature is commonly associated with vertebral rotation producing a rib hump which is most evident when the child stands in a forward bending position (Figure 20.1).

Scoliosis develops during growth (Cobb, 1948). It may be congenital or secondary to a neuromuscular or ligamentous disorder, but in most patients referred to hospital the cause is unknown (idiopathic). This last group develops as infantile (IIS), juvenile (JIS) or adolescent idiopathic scoliosis (AIS) named according to its age of onset. IIS affects slightly more boys than girls and is nearly always a thoracic curve convex to the left; most correct spontaneously, but of those that do not some become progressive curves which, without treatment, deform the child and produce cardiorespiratory compromise and death in early adult life (Wynne-Davies, 1975).

JIS and AIS affect girls predominantly (10 girls to one boy) (Ohtsuka et al., 1988). The scolioses are mainly right thoracic, right thoraco-lumbar, or left lumbar and some are double curves (James,

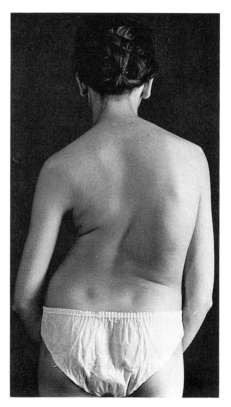

Figure 20.1 A girl with adolescent idiopathic scoliosis. Note the right thoracic scoliosis, the associated rib hump and the hip asymmetry.

1954). The side of the main curve is thought to be related to handedness. AIS occurs during the rapid growth of adolescence; peak height velocity (the turning point from growth acceleration to deceleration) is indicated about six months later by menarche. The detection of AIS has been aided in the last 20 years by school screening, particularly in the USA and Japan. In the UK, school screening for scoliosis is practised in a few centres, including Nottingham where the school nurses each year examine about 5000 children aged 12–13 years using a prescribed technique with a commercial inclinometer (Scoliometer). Subject to general practitioner approval they refer about 30 children each year to the Scoliosis Clinic. Overall, the scoliosis referrals to The Centre for Spinal Studies and Surgery at the University Hospital, Nottingham are mostly from general practitioners and consultants and include all types of scoliosis, idiopathic, congenital and secondary.

Previous Work on the Posterior Trunk Using Ultrasound

A few workers have used ultrasound to evaluate the shape of the spine and trunk in scoliosis. In Malmö, Sweden, Mauritzsen et al. (1985, 1986, 1991) employed an airborne ultrasound system to image the back surface in the standing position which measured asymmetry between right and left of the sagittal plane in the trunk. In Winnipeg, Canada, Letts et al. (1988) reported an ultrasonic digitization method for the measurement of scoliosis curves. In Newcastle, UK, Ions et al. (1986) and subsequently Oates et al. (1990) reconstructed postero-anterior (PA) and lateral views of the spine from segmental ultrasound scans in the prone position using a real-time linear array transducer. Suzuki et al. (1989) from Kyoto, Japan, using real-time ultrasound reported observations on laminal rotation in 47 patients with AIS, making comparisons with CT measurements of apical vertebral rotation in three patients. In Nottingham, ultrasound methods have been developed to appraise back shape in scoliosis since 1987 (Wojcik et al., 1988, 1990; Kiel et al., 1989).

TWO GROUPS OF PATIENTS IN THE NOTTINGHAM SCOLIOSIS CLINIC

With respect to the ultrasound evaluation two groups of patients with AIS are being studied:

1. Patients having surgical treatment, involving instrumentation and fusion.
2. Scoliosis school screening referrals.

SURFACE METHODS OF INVESTIGATING SCOLIOSIS

Integrated Shape Imaging System (ISIS)

This system is a TV-scanning method for appraising back shape and it is operated by the radiographer in the Department of Human Morphology in Nottingham. The patient stands and a line of light is projected down the patient's back. About 4000 data points are obtained each with x, y and z coordinates. The programme creates values for each of the following:

- lateral spinal curve angle;
- rib and lumbar hump in degrees (10 levels between T1 and S1);
- kyphosis and lordosis.

The procedure is then repeated with the patient in the sitting position.

Scoliometer

The scoliometer is a commercial inclinometer. When placed across the back of a patient with scoliosis in a forward bending position it gives an *angle of trunk inclination (ATI)* as an estimate of the size of the rib (and/or lumbar) hump. So as to compare scoliometer ATIs with the 10 level ISIS data for humps the scoliometer readings are obtained at 10 levels between T1 and S1, firstly in a standing, forward bending position and then in a sitting, forward bending position. Experience and attention to detail is essential in procuring ATIs using the scoliometer. There is a difference between the scoliometer ATIs and ISIS ATIs due to the position of the child; for ISIS, ATIs are obtained with the child standing upright and then sitting upright, while scoliometer ATIs are obtained in two forward bending positions (standing and sitting).

RADIOLOGICAL METHODS OF INVESTIGATING SCOLIOSIS

X-ray Films

The two standard projections of antero-posterior and lateral radiographs are obtained. In children proceeding to surgery lateral bending films are requested in the supine position. To appraise the effects of surgery on the spine and rib cage new techniques have been developed in Nottingham.

CT Scans

The aims of surgery are to correct the three-dimensional deformity of scoliosis and maintain it throughout life. The method of using segmental CT scans developed in Minneapolis is applied to patients before and after operation using established and new measurement techniques (Wood et al., 1991).

ULTRASOUND APPLICATIONS IN THE INVESTIGATION OF SCOLIOSIS

Back Shape

Currently *B-mode ultrasound* is used to assess back shape at 10 levels between T1 and S1 with the patient in a prone position (to compare with each of ISIS and scoliometer ATI data). The hard copies of the ultrasonograms enable spinal rotation and rib rotation to be calculated relative to the sacrum. This type of ultrasound examination has been in use since 1990. In patients having surgery, the examinations are performed before and, at repeated intervals up to two years, after instrumentation and spinal fusion. In the school screening referrals, the examinations are performed at intervals according to the severity of the scoliosis curve. Most recently, new methods to measure spinal rotation and rib rotation have been developed using a *real-time* portable machine with a long linear array transducer.

Torsion in Lower Limb Bones

Femoral anteversion and tibial torsion are bony axial rotations which diminish in the femur and increase externally in the tibia as a normal child grows from birth to adolescence. Normal asymmetries of femoral anteversion and tibial torsion occur. In 1988, a B-mode method was developed in Mansfield and Nottingham for measuring femoral anteversion (Upadhyay et al., 1987a, 1987b). It was applied to preoperative patients with scoliosis, school screening referrals and their siblings and the findings compared with healthy controls (Burwell et al., 1988; Upadhyay et al., 1990). The findings helped to initiate a new concept of scoliosis causation (Burwell et al., 1989; Burwell and Dangerfield, 1992; Burwell, 1994).

Since 1993, real-time ultrasound methods have been developed to evaluate both femoral anteversion and tibial torsion in patients with lower limb disorders and have also been applied to patients with AIS to help to elucidate its causation. These ultrasound methods have been applied to both surgical patients and screening referrals. All this research has necessitated the collection of data on femoral anteversion and tibial torsion from 520 healthy children aged 5–18 years in schools in order to provide a database to compare with the patient groups (Kirby et al., 1996).

ULTRASOUND METHODS

Back Shape

Static B-mode

The back is marked into 10 levels from T1 to S1 (to compare scoliometer ATI readings with the ISIS ATI readings). These same 10 marks are used for the ultrasound images. A static B-mode ultrasound system

(Technicare EDP 1200) is used to image spinal and rib rotation, and the hard copy films are obtained with a Matrix single format video imager (Figure 20.2). A 7.5 MHz transducer is used unless the child is obese. The patient lies prone with the forehead supported on a special tripod with the arms hanging over the top of the examination couch to clear the scapulae from the ribs. The gantry is aligned with the spine which allows travel of the transducer arm from T1 to S1 without further adjustment. The first image is obtained at the level of S1 and includes the sacral grooves and each ilium. At higher spinal levels, images are obtained of the spinous processes and posterior aspects of the laminae, including the ribs. From the hard copies *spinal rotation* is measured as the angle between a line drawn through the vertebral laminae and the horizontal (Figure 20.3a); and *rib rotation* as the angle between a line joining the most posterior aspects of the ribs and the horizontal (Figure 20.3b). These angles are then corrected relative to the sacrum.

Real-time portable ultrasound machine with long transducer

The temporary availability of a portable ultrasound machine (Aloka SSD 500) with a veterinary linear array transducer recently provided the opportunity for a feasibility study of a new ultrasound method for back shape appraisal for scoliosis. Readings of spinal and rib rotation were made at five levels on the back with each child in the standard prone position.

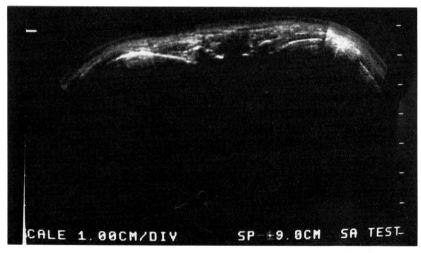

Figure 20.2 Ultrasonogram of the posterior chest wall of a patient with right thoracic scoliosis. Note the laminae of the vertebra and the arc of the ribs.

Ultrasound and Scoliosis 243

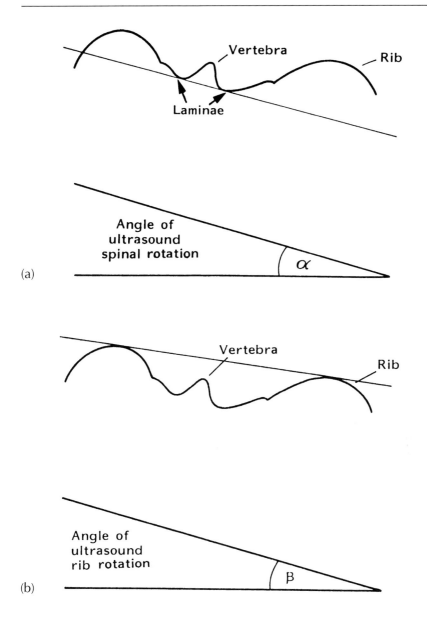

Figure 20.3 (a): Method of calculating the angle of spinal rotation (α) relative to the sacrum. (b): Method of calculating rib rotation (β) relative to the sacrum. (Modified from Burwell et al., 1988.)

Femoral Anteversion and Tibial Torsion

These are measured using a real-time ultrasound technique. An Aloka SSD 650 machine and a wide view 3.5 MHz linear array transducer are used together with a gravity inclinometer to measure angles. The patient is supine with the lower limbs free from the examination couch and the ankles supported on a separate plinth. The latter is set slightly higher than the couch to ensure complete extension of the knees.

Tibial torsion (Kirby et al., 1994)

With the transducer inverted and maintained in a horizontal position, as indicated by the inclinometer attached to the probe, the posterior aspect of the tibial condyles is imaged just distal to the tibial plateau. The lower limb is rotated until this image appears horizontal on the monitor. Immobilizing the limb in this position allows the transducer to be removed and replaced on the front of the ankle to image the trochlear surface of the talus. By rotating the transducer to produce a horizontal image on the monitor, the angle of rotation is read from the inclinometer to measure the angle of tibial torsion, or more strictly the angle of tibio-talar torsion. (There is little axial rotation of the talus in the ankle joint mortice.)

Femoral anteversion (Zarate et al., 1983)

This technique is repeated for the femur. The image of the posterior femoral condyles is made horizontal by rotating the limb; then the probe is used to image the head of femur and greater trochanter; by rotating the probe until this image is horizontal, the angle of femoral anteversion is read from the inclinometer attached to the probe.

SUMMARY OF ULTRASOUND FINDINGS

Back Shape

Reliability of the methods

In 10 patients scanned twice and moving around the room between each set of scans, the reliability of one observer measuring spinal rotation and rib rotation relative to the sacrum was ± 3° (95% confidence limits).

Pre- and postoperative patients

Preliminary observations on 10 patients with AIS having anterior spinal surgery (Zielke procedure) were examined using the B-mode ultrasound method preoperatively, postoperatively and up to two years (Whitwell et al., 1994, 1995a). All the patients showed initial diminution of spinal rotation and most a diminution also of rib rotation. Two patients with lumbar curves showed an exaggeration of rib rotation after the operation. In one patient after Zielke procedure the derotation of the spine was transmitted to the pelvis. In follow-up beyond one year, three patients showed a reassertion of rib rotation with or without an increase in spinal rotation. It will be appreciated that this ultrasound method provides information about rib rotation (rib hump) which is not seen on conventional AP and lateral spinal radiographs. Axial CT scans of the trunk show rib rotation (in the hump), but it exposes the patient to additional X-rays and is expensive.

School screening referrals

1. *Ultrasound rib rotation as a predictor of spinal rotation.* In 100 children referred after school screening for scoliosis it was shown that ultrasound rib rotation is a better predictor of spinal rotation than either of the surface methods for measuring the angle of trunk inclination (ISIS and Scoliometer, Polak et al., 1995).
2. *Ultrasound rib rotation and spinal rotation.* In 100 children referred after school screening for scoliosis and in 41 preoperative patients, calculations were made to examine the relationship of spinal rotation to rib rotation, by calculating the position of equal axial rotation (PEAR) (Whitwell et al., 1995b). The findings suggest the role of rib rotation (muscular control) as well as vertebral factors (adaptive growth) in the pathogenesis of AIS.
3. *Preliminary evaluation of a real-time portable ultrasound machine and a long transducer.* In eight school screening referrals, the intraobserver reproducibility of this method for each of spinal rotation and rib rotation was ± 1.8° to ± 3.6° (95% confidence limits, Kirby et al., 1995).

Femoral Anteversion and Tibial Torsion

Femoral anteversion and tibial torsion

Preliminary observations using B-mode to measure femoral anteversion in patients with scoliosis and their siblings have been reported

(Burwell et al., 1988, 1989). More recently, preliminary observations using real-time ultrasound to evaluate femoral anteversion and tibial torsion in preoperative patients and school screening referrals have been reported (Burwell et al., 1996a,b). More patients are needed in this research which is continuing.

Relation of surface back shape to torsion in lower limb bones

Preliminary observations in 63 school screening referrals and operative patients with AIS suggest a relation between back shape and the torsional patterns in femur and tibia (Cavdar et al., 1995). These observations are consistent with the Nottingham concept of causation of AIS.

DISCUSSION

The rapid advance in spinal surgery for scoliosis

In the last 10 years much progress has been made in the metallic internal fixation used to correct the spinal deformity of scoliosis and to maintain it; the internal fixation is so rigid that external braces or plasters are no longer needed after surgery. The results of these new surgical methods need appraising and the ultrasound work reported here has been developed, among other methods, for this purpose.

Research or service?

The ultrasound work is performed as research and not as part of normal NHS practice. The aims are threefold: firstly, to evaluate the effectiveness of the new spinal instrumentations for scoliosis; secondly, to improve screening for scoliosis, and thirdly, to understand the causation of idiopathic scoliosis.

Future

The preliminary research suggests that ultrasound will prove to be the best method for the appraisal of back shape of patients with scoliosis. A portable real-time ultrasound machine with a long probe holds promise of a simple method which could be used by radiographers who are involved in this type of surgery or in screening for AIS. In the longterm, surgery for scoliosis, which currently is directed at straightening a bent spine, should ultimately be based on some knowledge of the causation of AIS. The ultrasound research is contributing to this end.

Acknowledgements

The writer is grateful to Mr J. K. Webb FRCS, Director of The Centre for Spinal Studies and Surgery, University Hospital, Nottingham, for the opportunity to examine patients under his care, to Emeritus Professor R. G. Burwell for support, and to Professor T.M. Mayhew for some facilities.

The research discussed in this chapter would not be possible without teamwork and the support of the funding bodies. The core of the work is the management of the patients with scoliosis under the care of Mr J. K. Webb FRCS, Consultant Spinal Surgeon. The research staff in addition to a Research Radiographer (Mrs A. S. Kirby MSc DCRR) includes a Research Professor (Emeritus Professor R. G. Burwell), a Spinal Research Fellow (formerly Dr A. A. Cole and currently Mr R. K. Pratt FRCS), a Research Physiotherapist (formerly Mrs F. J. Polak MSc MCSP) and a secretary. The research was funded by Action Research until 1993 and subsequently by AO/ASIF of Switzerland. The Arthritis and Rheumatism Council of the UK is funding the study of femoral anteversion and tibial torsion in healthy subjects.

References

Burwell, R.G. (1994) The aetiology and pathogenesis of adolescent idiopathic scoliosis (AIS). Summary of a 3D multifactorial concept. *European Spinal Resonances* **4:** 3–6.

Burwell, R.G., Dangerfield, P.H. (1992) The pathogenesis and assessment of idiopathic scoliosis. In Findlay, G., Owen, R. (eds) *Surgery of the Spine, a Combined Orthopaedic and Neurosurgical Approach*, Chapter 19, Vol. 1, pp. 365–408. Oxford: Blackwell Scientific Publications.

Burwell, R.G., Upadhyay, S.S., Wojcik, A.S. et al. (1988) Ultrasound in evaluating scoliosis. In Siegler, D., Harrison, D., Edgar, M. (eds) *Prognosis in Scoliosis:* Proceedings of the Eighth Phillip Zorab Scoliosis Symposium, London, October 1988, pp. 49–67. British Scoliosis Research Foundation.

Burwell, R.G., Upadhyay, S.S., Webb, J.K. et al. (1989) Femoral anteversion and the declive angle in 12–13 year old children referred by new screening criteria. A new theory of aetiology. *Journal of Pediatric Orthopedics* **9:** 349.

Burwell, R.G., Kirby, A.S., Moulton, A. (1996a) Torsion in lower limb bones of children screened for adolescent idiopathic scoliosis (AIS). In *Proceedings of the International Research Society of Spinal Deformities*. First meeting, Stockholm, June 16–19, 1996.

Burwell, R.G., Kirby, A.S., Cole, A.A. (1996b) Torsion in lower limb bones of patients with adolescent idiopathic scoliosis (AIS) treated surgically. In *Proceedings of the International Research Society of Spinal Deformities*. First meeting, Stockholm, June 16–19, 1996.

Cavdar, S., Burwell, R.G., Cole, A.A. et al. (1995) The relation between femoral anteversion, tibial torsion and back shape in children with adolescent idiopathic scoliosis (AIS). *Clinical Anatomy* **8:** 367.

Cobb, J.R. (1948) *Outline for Study of Scoliosis*. American Academy of Orthopaedic Surgeons Instructional Course Lectures, pp. 261–275. St. Louis: CV Mosby.

Ions, G.K., Whittingham, T.A., Saunders. P.J. et al. (1986). Radiation free imaging of the scoliotic spine. In Harris, J.D., Turner-Smith, A.R. (eds) *Surface Topography and Spinal Deformity: Proceedings of the 3rd International Symposium*, Oxford, 27–28 September 1984, pp. 71–75. Stuttgart: Gustav Fischer.

James, J.I.P. (1954) Idiopathic scoliosis; the prognosis, diagnosis and operative indications related to curve patterns and the age of onset. *Journal of Bone and Joint Surgery* **36B:** 36–49.

Kiel, A.W., Burwell, R.G., Moulton, A. et al. (1989) The spinal declive angle: an appraisal using surface, radiological and ultrasound methods. In: Scoliosis Research Society combined meeting with European Spinal Deformities Society, 17–22 September, Amsterdam, Netherlands.

Kirby, A.S., Burwell, R.G., Aujla, R.K. et al. (1994) A new ultrasound method for measuring tibial torsion (TT) angles and findings in young adolescents. *Clinical Anatomy* **7:** 56.

Kirby, A.S., Aujla, R.K., Burwell, R.G. et al. (1995) A preliminary study of a new real-time ultrasound method for measuring rib and spinal rotation in scoliosis. *Clinical Anatomy* **8:** 369.

Kirby, A.S., Aujla, R.K., Burwell, R.G. et al. (1996) A cross-sectional ultrasound study of femoral anteversion and tibial torsion in healthy children aged 5–17 years. *Clinical Anatomy* (in press).

Letts, M., Quanbury, A., Gouw, G. et al. (1988) Computerized ultrasonic digitization in the measurement of spinal curvature. *Spine* **13:** 1106–1110.

Mauritzsen, L., Benoni, G., Lindström, K., Willner, S. (1985) Imaging the form of the back with airborne ultrasound. In Whittle, M., Harris, D. (eds) *Biomechanical Measurement in Orthopaedic Practice*, Chapter 11, pp. 87–91. Oxford: Clarendon Press.

Mauritzsen, L., Benoni, G., Ilver, J. et al. (1986) Three linear array back scanning with airborne ultrasound. In Harris, J.D., Turner-Smith, A.R. (eds) *Surface Topography and Spinal Deformity: Proceedings of the 3rd International Symposium*, Oxford, 27–28 September 1984, pp. 47–51. Stuttgart: Gustav Fischer.

Mauritzsen, L., Ilver, J., Benoni, G. et al. (1991) Two-dimensional airborne ultrasound real-time linear array scanner applied to screening for scoliosis. *Ultrasound in Medicine and Biology* **17(5):** 519–528.

Oates, C.P., Whittingham, T.A., Leonard, M.A. (1990) The Newcastle spine imaging system. In Neugebauer, H., Windischbauer, G. (eds) *Surface Topography and Spinal Deformity: Proceedings of the 5th International Symposium*, Vienna, Austria, 29 September to 1 October 1988, pp. 31–33. Stuttgart: Gustav Fischer.

Ohtsuka, Y., Yamagata, M., Arai, S. et al. (1988) School screening for scoliosis by the Chiba University Medical School Screening Programme. Results of 1.24 million students over an 8-year period. *Spine* **13:** 1251–1257.

Polak, F.J., Whitwell, D.J., Burwell, R.G. et al. (1995) Prediction of spinal rotation in school screening referrals for adolescent idiopathic scoliosis: the advantage of ultrasound over surface methods. *Clinical Anatomy* **8**: 152 and *Journal of Bone and Joint Surgery Orthopaedic Proceedings* **77B**: 258.

Suzuki, S., Yamamuro, T., Shikata, J. et al. (1989) Ultrasound measurement of vertebral rotation in idiopathic scoliosis. *Journal of Bone and Joint Surgery* **71B**: 252–255.

Upadhyay, S.S., O'Neil, T., Burwell, R.G., Moulton, A. (1987a) A new method using medical ultrasound for measuring femoral anteversion (torsion): technique and reliability. *British Journal of Radiology* **60**: 519–523.

Upadhyay, S.S., O'Neil, T., Burwell, R.G., Moulton, A. (1987b) A new method using ultrasound for measuring femoral anteversion (torsion): technique and reliability. An intra-observer and inter-observer study on dried bones from human adults. *Journal of Anatomy (London)* **155**: 119–132.

Upadhyay, S.S., Burwell, R.G., Moulton, A. et al. (1990) Femoral anteversion in healthy children. Application of a new method using ultrasound. *Journal of Anatomy (London)* **169**: 49–61.

Whitwell, D.J., Webb, J.K., Burwell, R.G. et al. (1994) Zielke VDS: findings obtained by a multilevel ultrasound method for measuring spinal and rib torsion compared with surface, radiological and CT methods. *Journal of Bone and Joint Surgery, Orthopaedic Proceedings* **76B**: 9.

Whitwell, D.J., Kirby, A.S., Burwell, R.G. et al. (1995a) Ultrasound evaluation of spinal and rib rotation in preoperative patients with adolescent idiopathic scoliosis (AIS) in relation to the surface deformity, with observations on Zielke VDS. *Journal of Bone and Joint Surgery, Orthopaedic Proceedings* **77B**: 31.

Whitwell, D.J., Kirby, A.S., Burwell, R.G. et al. (1995b) Ultrasound rib and vertebral rotation in the trunk of screening referrals and pre-operative patients with adolescent idiopathic scoliosis (AIS). *Clinical Anatomy* **8**: 153.

Wojcik, A.S., Burwell, R.G., Webb, J.K., Moulton, A. (1988) A new ultrasound method for measuring spinal torsion in scoliosis. *Journal of Bone and Joint Surgery* **71B**: 728.

Wojcik, A.S., Burwell, R.G., Webb, J.K., Moulton, A. (1990) A new ultrasound method for measuring spinal torsion in scoliosis. In Neugebauer, H., Windischbauer, G. (eds) *Surface Topography and Spinal Deformity: Proceedings of the 5th International Symposium*, Vienna, Austria, 29 September to 1 October 1988, pp. 25–30. Stuttgart: Gustav Fischer.

Wood, K.B., Transfeldt, E.E., Ogilvie, J.W. et al. (1991) Rotational changes of the vertebral-pelvic axis following Cotrel-Dubousset instrumentation. *Spine* **16**: 404–408.

Wynne-Davies, R. (1975) Infantile idiopathic scoliosis. Causative factors, particularly in the first six months of life. *Journal of Bone and Joint Surgery* **57B**: 138–141.

Zarate, R., Cuny, C., Sazos, P. (1983) Détermination de l'antéversion du col du fémur par échographie. *Journal of Radiology* **64**: 307–311.

21
Telemedicine and Teleradiology in Primary Care: a GP's Perspective

John Wynn-Jones and Susan Groves-Phillips

INTRODUCTION

Telemedicine has been defined (AIM, 1990) as 'the investigation, monitoring and management of patients and the education of patients and staff using systems which allow ready access to expert advice and patient information no matter where the patient or relevant information is located'. Telemedicine has an important role in reducing the inequalities of access to health care experienced by GPs and patients in rural communities. Patients often have to travel significant distances to receive specialist treatment at district general hospitals. The real benefit of telemedicine is that it will reduce these distances, bringing specialist care closer to the patient. Preston et al. (1992) wrote 'Telemedicine puts the specialist where the specialist is needed. It brings the medical educator and primary health care provider to the electronic examining table.' Telemedicine has the added benefit of enabling the primary health care clinician to draw upon a huge resource of on line expert knowledge whenever and wherever it is needed.

Teleradiology is one of the applications of telemedicine where considerable research and development has already been carried out. Much of this is based in the more remote areas of the developed world where 'images are sent to radiologists who are linked to each other, allowing dialogue and the dissemination of skills and knowledge over vast areas', Viitanen et al. (1992). 'Teleradiology in the USA has become part of everyday practice' (Wright and Loughrey, 1995). It is difficult to give a general practitioner's (GP's) perspective on teleradiology as it has yet to make an impact upon their working lives, but the potential exists to improve the provision and quality of primary health care. It will facilitate the transfer of skills from the radiologist to the GP and from secondary into primary care.

In order to give a GP's perspective on teleradiology, and to a lesser extent on telemedicine it is important that the following issues are addressed:

- The need for imaging and teleradiology in primary care in the UK.
- How far have these needs been met?
- The advantages and disadvantages of teleradiology.
- Conclusions.

THE NEED FOR IMAGING AND TELERADIOLOGY IN PRIMARY CARE IN THE UK

An audit of waiting times for X-rays and associated imaging results in a practice in rural mid-Wales was carried out. It showed that the average waiting time for the return of imaging results was seven days. The longest waiting time for a result was 26 days. A request for a barium study on average, takes 24 days before the patient is seen. These lengthy waiting times are a cause for anxiety to patients. The delays are due to the problems of rurality and distance related to travelling to the nearest x-ray department and waiting for the images to be seen by a radiologist able to give an expert opinion. The Welsh Health Planning Forum (1994a) describes the problems faced by community hospitals providing x-ray services in Wales: 'A significant number of x-ray films have to be sent by road to another site for reporting, the report and the films are then sent back in the same manner'. The delays in the turn-around of results from imaging requests would be reduced, by the use of a teleradiology system, as images would be much quicker to process and could be sent back together with the report through a teleradiology link to the requesting GP for consideration. The Welsh Health Planning Forum (1994a) lists the claimed benefits of using x-ray images in digital form (teleradiology):

- Images captured digitally when displayed can be manipulated and altered.

- There is a reduction in the radiation dose a patient receives. (This claim has not as yet been proven.)
- Films are less likely to be mislaid and the need to repeat them removed.
- Savings in staff, space and consumables.
- Transfer of images is easier (via teleradiology).

The digitization of images may have a wide impact upon the use made of imaging techniques by GPs. At present x-ray films and other images are owned by the trust or authority from whom they are commissioned. GPs are given limited access to images. The ability to see the radiograph and compare the report with the image would be beneficial to the GP's understanding of the imaging process, and the appropriateness of using different imaging techniques.

In the report *Health and Social Care 2010: A Report of Phase One* the Welsh Health Planning Forum (1994b) states that 'the volume of minimal access surgery will increase to 80% by the year 2000'. This increase in minimal access surgery will stimulate the need for more sophisticated imaging techniques and the report predicts that over the next few years the overall number of x-ray investigations will not increase and may even fall, giving way to other imaging techniques. It is possible to predict that in the future, a wider range of health care professionals will be using x-ray imaging, and x-ray equipment will use digital technology linking through to established health care networks. Teleradiology will facilitate the changes needed in the infrastructure of the NHS to allow for the re-siting of x-ray departments in new locations, to provide the best access for the patient. It will also support the backup needed for diagnostic imaging and interventional radiology, and will lead to the 'prudent' evaluation of the costs and benefits of digitization and networking.

'Many rural communities are serviced by peripatetic radiologists travelling great distances in order to perform contrast and screening studies' Welsh Health Planning Forum (1994a). It would be much more efficient to have trained radiographers carrying out these tasks with on-line radiologists to give opinions when required.

HOW FAR HAVE THESE NEEDS BEEN MET?

At St Mary's Hospital on the Isle of Wight the obstetric ultrasound unit is linked via an Integrated Services Digital Network (ISDN) link to the Centre for Fetal Care at Queen Charlotte's & Chelsea Hospital in London, 'The aim of this link is to provide an almost instantaneous expert opinion, within about 15–30 minutes of the request being made' Fisk et al. (1995). Before this link was available patients were sent to London to be seen by a specialist in Queen Charlotte's & Chelsea Hospital when any irregularities were found during ultrasound examinations. This caused a great deal of inconvenience and stress to the patient. The findings from the study of this link are that 'The application of telemedicine in this field appears to have several advantages over current practice in the National Health Service' (Fisk et al., 1995).

The Welsh Health Planning Forum (1994a) listed the requirements for the creation of long distance links as:

- Digitizing facilities at the hospital performing the investigation.
- Image-receiving equipment and high resolution viewing monitors at the receiving site.
- A network capable of rapidly transferring large amounts of data.
- Staff available to report on the image at the receiving site.
- Arrangements to transfer the report back to the community hospital and/or GP.
- Clear consistent standards for data systems and communications.

The cost of digitizing equipment falls as the sophistication grows. New radiology systems use digital technology such as 'The PACS (Picture Archiving and Communications System) at The Hammersmith Hospital where there are 138 workstations distributed throughout the hospital and medical school, each linked to the central image store' (Glass et al., 1993). The system utilizes reusable photo stimulable phosper plates to produce digital X-ray images. PACS allows for multisite access and sharing of the same image. The adoption of similar systems in radiology departments could aid the development of shared electronic record systems, with GPs using ISDN links to access hospital networks in order to receive information about their patients from PACS systems, pathology departments and other Hospital Information Systems.

The ability to link together different digital equipment at different sites is essential to the success of teleradiology and standards are being agreed to ensure that this can be done. The DICOM standard ensures that images can be exported in a common format acceptable by all conformant equipment. There are standards for visual systems and the equipment for teleconferencing known as the family of ITU H.320 standards. Video coding standards are covered by H.261.

Midwives use portable ultrasound scanners to monitor their patients in urban GP surgeries, as well as in rural and remote areas. The interpretation of ultrasound images is difficult, but a direct teleradiology link could provide on-line expert help to interpret the images and guide the practitioner to obtain the best results. There is a growing trend towards a more widespread adoption of ultrasound techniques by GPs. Ultrasound currently accounts for 15% of all imaging procedures. Ultrasound expansion in primary care has been fuelled by resources released by fundholding. This has also been facilitated by the reduction in cost, increase in sophistication and the portability of equipment. The Welsh Health Planning Forum (1994a) see this trend as one which will continue and will make ultrasound second only to X-rays in use. At present the main application of ultrasound is for monitoring fetal development, but its range of use is growing.

Traditionally the x-ray service has been based in the secondary care sector. This is mainly due to the nature and structure of the NHS, the problems associated with health and safety risks from medical radiation exposure, the cost of purchasing the necessary equipment and the setting up of suitable premises. Another inhibiting factor appears to be the system for payment and reimbursement of GPs in the UK. This restricts service development by denying the GP the option of an 'Item for service charge' to cover their capital and running costs. In veterinary medicine and in dentistry, x-ray equipment is available in most surgeries. Imaging facilities within GP surgeries is common in countries such as Germany and the USA where GPs are reimbursed for their work. Very few GPs in the UK have ventured into this area. It is probable that although some GPs, particularly fundholders, may eventually invest in their own radiology equipment, the greatest impact teleradiology will have on GPs will be in the community hospital environment, where rural GPs in the UK often provide medical cover for casualty departments, ordering and acting on radiographs in acute situations. Expert on-line back-up via telemedicine and teleradiology, in this environment would prove invaluable.

The growing primary health care team consists of an ever expanding range of professionals, all with new skills, expectations and training needs. Some are based in the GP surgery while others in rural areas work from community hospitals. It is difficult to gauge the level of enthusiasm for technological innovations amongst these professionals. Most, however, agree that change is inevitable.

THE ADVANTAGES AND DISADVANTAGES OF TELERADIOLOGY

To date, the impact of imaging on the average GP has been limited. Most GPs now have access to radiology departments for basic diagnostic tests, but it is still true that the provision and management of imaging is based in and designed for secondary and tertiary care. The incorporation of images into an electronic patient record, accessible from any health care site where the patient receives treatment, would require an investment in training. This investment would, however, have a beneficial impact upon patient care as GPs would be able to assess the progress of chronic conditions and improve patient education by allowing them to view their own results.

If these diagnostic services become an integral part of primary care, the shift in emphasis will have large implications for the provision, planning and costing of the radiology services in the future. Advanced telematics has, and will have, an impact on the delivery, efficiency and safety of such a service. The benefits of teleradiology may be best appreciated in the area of acute services. Small community hospitals treat minor injuries and act as triage and resuscitation units for more serious conditions. The ability to send images to trauma units from these smaller community units will enhance the ability of these practitioners to provide a more comprehensive service and at the same time will save patients from travelling long distances to large centralized hospitals. It is in this setting where GPs commission and take radiographs, that the need for expert support to interpret images is greatest. GP stress and isolation could be reduced by contact with colleagues and patient care improved through the linking of hospitals and GPs into a health service network.

The scope for teleradiology is expanding but problems still exist. Despite the many possible advantages, it is important to declare a note of caution. The expansion of an imaging service may lead to less central and professional control. The volume of work carried out in small units and surgeries may not be adequate to maintain skills and standards. The possibilities for misinterpretation are present, leading to possible tragedy.

The compression of digital images for transmission can lead to some loss of quality (Hartviksen and Rinde, 1993). 'Spatial resolution may be less with digital than analogue images' (Greene and Oestmann, 1992) leading to problems in chest and skeletal radiology which can tolerate less compression. 'The high resolution required makes mammography unsuitable at present for transmission' (Wright and Loughrey, 1995).

Digitized images are satisfactory for some standard x-ray projections, ultrasound, fluoroscopy, computed tomography and magnetic resonance imaging. It has to be pointed out that the technology of transmitting these images is, however, constantly improving.

The ease of communication may lead to the closure of small departments and the centralization of services in large centres, remote from GPs and primary care. Capital costs may take funds away from other essential patient care as community hospitals feel that their survival may depend on the investment in extensive technology. GP expansion may also result in the removal of services from the community hospital making radiology in community hospitals even less viable. User acceptance of these new technologies can also prove a problem as this technology may be perceived as reducing the need for radiologists in the future.

Wright and Loughrey (1995) stated that technology is racing ahead of regulators in the USA. Professional bodies, government organizations and defence bodies have been slow to match the advances made in telemedicine. It is hoped that regulation will be governed by precedent rather than vision.

CONCLUSIONS

Some of the developments discussed above are already possible while others depend on a certain degree of vision. Cost may prohibit the widespread uptake of x-ray equipment in GP surgeries, but it is probable that there will be a growth of teleradiology linked to community hospitals. GPs working in these hospitals may soon feel that they must have direct access to opinions and management advice for medico-legal reasons. The further uptake and developments in the fields of scanners, reusable plates, and digital links may make access to centres of excellence not only desirable but necessary. Image storage and access problems are being addressed as data storage capacities expand and health networks are set up. Education and training will be of the utmost importance as will the setting of standards and guidelines.

Many changes will become possible as technology develops. Digital radiology images will be stored as data files becoming an integrated part of truly transferable patient electronic records via a distributed database system connecting all health care sites together and allowing the electronic record to be accessible from wherever the patient is receiving treatment. It is also possible that there will be the development of centres of excellence providing 24 hour on-line support for GPs and reporting on images at a distance. Major teaching hospitals hit by financial pressures may see this as a solution for maintaining staff and expanding services. Remote screening centres could be set up by specially trained radiographers providing access for remote populations to services such as echocardiography and barium studies. These centres could be serviced and supported by specialist centres sited anywhere in the UK or the rest of Europe.

The future provides threats and challenges for all and it is necessary to be committed to the principle that only improvements to patient care should drive change. Needs rather than technology should drive research. The expansion of imaging in primary care should lead to efficiency in the use of resources; care based on clinical effectiveness, and be of direct benefit to the public. The National Health Service in recent years has undergone wide-ranging reforms which have changed the emphasis from secondary to primary care. Planning is increasingly based upon the local needs of the patients. Fundholding has empowered GPs to provide a wider range of services outside their traditional remit. These changes have led to the raising of both professional and lay expectations, leading to the closer scrutiny of how and by whom health care is provided.

References

Advanced Informatics in Medicine (AIM) (1990) Supplement Application of Telecommunications of Health Care Telemedicine. AI 1685.

Fisk, N.M., Bower, S., Sepulveda, W. et al. (1995) Fetal telemedicine: interactive transfer of real-time ultrasound and video via ISDN for remote consultation. *Journal of Telemedicine and Telecare* **1**: 38–44.

Glass, H., Reynolds, R.A., Allison, D.J. (1993) Planning for PACS at Hammersmith Hospital. *Proceedings of the Nordic Symposium on PACS, Digital Radiology and Telemedicine*, pp. 37–50.

Greene, R.E., Oestmann, J.W. (1992) The rationale for transition to digital radiography. *Computed Digital Radiography in Clinical Practice*. New York: Thieme Medical.

Hartviksen, G., Rinde, E. (1993) Telecommunication for remote consultation and diagnoses. *Telektronikk* **1**: 23–32.

Peredia, D.A., Allen, A. (1995) Telemedicine technology and clinical applications. *Journal of the American Medical Association* **273(6)**: 483–488.

Preston, J., Brown, F.W., Hartley, B. (1992) Using telemedicine to improve health care in distant areas. *Hospital and Community Psychiatry* **43(1)**: 25–32.

Royal College of Radiologists (1993) *Making the Best Use of a Department of Clinical Radiology, Guidelines for Doctors*. London: Royal College of Radiologists.

Viitanen, J. et al. (1992) Nordic teleradiology development. *Computer Methods and Programs in Biomedicine* **37:** 273–277.

Welsh Health Planning Forum (1994a) *Getting the Best for Patients, Effectiveness in Medical Imaging*.

Welsh Health Planning Forum (1994b) *Health and Social Care 2010: A Report of Phase One*.

Wright, R., Loughrey, C. (1995) Teleradiology. *British Medical Journal* **310:** 1392–1393.

22
Current Trends in Nuclear Medicine

Daphne Glass and A. Michael Peters

Advances in nuclear medicine have been based broadly on three fundamental areas: molecular and cell biology, radiochemistry and physics. Additionally, system-specific advances have been made, including the imaging of inflammation, the cardiovascular system and in oncology. This chapter addresses these advances and some of the challenges they provide for those involved in nuclear medicine.

MOLECULAR AND CELL BIOLOGY

Tracers in routine use in nuclear medicine are generally single elements or simple compounds, which tend to have high sensitivity but a low specificity. Examples of single element tracers include radio-iodine for thyroid investigations, inert noble gases, such as krypton-81m (Kr-81m) and xenon-133 (Xe-133) for ventilation or blood flow studies, and fluorine-18 (F-18) for positron emission tomography (PET) of the skeleton. These tracers have been in use since the beginnings of nuclear medicine and preceded the development of the gamma camera.

After the introduction of the gamma camera and the development of technetium-99m (Tc-99m) chemistry, simple compounds were introduced. These were designed chemically to target specific tissues following intravenous injection. Examples of these are Tc-99m methylene disphosphonate (MDP) and Tc-99m dimercaptosuccinic acid (DMSA).

Tc-99m diethyltriaminepentaacetic acid (DTPA) is similar although it has less tissue specificity than the first two and can be regarded as nuclear medicine's utility tracer – a sort of nuclear medicine contrast agent.

This phase in the development of tracers was followed by the introduction of 'designer' molecules with greater specificity in the sense that they were designed to target specific cells. In contrast, although DMSA and MDP target kidney and bone respectively, their precise modes of uptake remain unclear. Radio-iodinated meta-iodobenzylguanidine (MIBG) is a good example of an agent targeted at a specific and well characterized physiological system – it substitutes for noradrenaline in the post-synaptic sympathomimetic amine cycle. More recently, tracers have been designed to target specific subcellular structures. Thus indium-111 (In-111)-labelled octreotide, an analogue of somatostatin, targets the somatostatin receptor which is expressed in a variety of neoplastic and neoplastic-like conditions (Figure 22.1). This was one of the first agents, apart from a monoclonal antibody, developed to label a specific receptor (Lamberts et al., 1990). All these examples are small

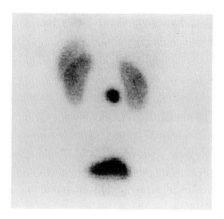

Figure 22.1 In-111 octreotide accumulation in a duodenal gastrinoma.

molecules, which are small enough to diffuse readily across the endothelium and undergo glomerular filtration. Large designer molecules have also been introduced, of which the best examples are probably the monoclonal antibodies.

The designer element of these is self-evident but a more subtle large molecule designed for a specific purpose is serum amyloid P component

(SAP) labelled with iodine-123 (I-123) (Hawkins et al., 1988) (Figure 22.2). The introduction of this agent into nuclear medicine was a remarkable achievement as the molecule is a natural plasma protein and, yet, targets all known forms of amyloid deposit. It has proved very effective for imaging amyloid deposits in the liver, spleen, bone marrow, kidneys and adrenal glands but has been disappointing in the detection of amyloidosis of skeletal muscle, myocardium and skin, and in the cerebral amyloidosis associated with Alzheimer's disease. This may be the result of its size: these tissues have a continuous endothelium in contrast to the fenestrated endothelium of the reticuloendothelial system and adrenal glands and, like the monoclonal antibodies, I-123-SAP is likely to have an access problem in these tissues.

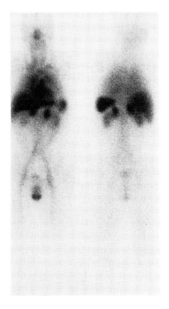

Figure 22.2 Abnormal I-123 labelled serum amyloid P component (SAP) accumulation in the spleen and adrenal glands.

Another category of agent is the living cell. When cell labelling was developed in the late seventies, it seemed an esoteric and ambitious aim, but white cell scanning for infection and inflammation has become one of the most valuable and frequently used techniques in nuclear medicine (Peters, 1994). The designer molecules are based on principles of molecular biology, while cell labelling is based on cell biology. These two disciplines of molecular medicine, molecular and cell biology, are, perhaps, the most rapidly expanding areas in medicine and it is clear that advances in nuclear medicine are critically dependent on exploiting these areas. With radiochemistry as sophisticated as it now is, it is reasonable to suggest that almost all biological mediators, including receptor ligands, could be labelled with gamma-emitting radionuclides and their target molecules imaged. All cells of the body are communicating constantly through these molecules. The challenge facing nuclear medicine is to identify the individual conversations; analogous to isolating and listening to two people talking to each other in a full sports stadium against the background of the thousands of other simultaneous conversations.

RADIOCHEMISTRY

Exploitation of the explosion of knowledge in molecular medicine requires parallel advances in radiochemistry. This is already sophisticated but needs to become even more so. A fundamental problem in physiological imaging is that the chemical manipulation required to label a compound alters its biological properties as a result of the insertion of a foreign element. This is one reason why positron-emitting radiopharmaceuticals are so elegant, since labelling with carbon-11 (C-11), for example, replaces one carbon atom with another, thereby leaving the chemical properties of the compound completely unaltered. Placing an iodine or technetium molecule into a delicate biological mediator is, however, likely to disturb its function.

A recent intriguing development has been the engineering by genetic manipulation of antibody fragments to contain peptide tails of a sequence specifically designed to bind Tc-99m (George et al., 1995). This approach combines advances in molecular biology with advances in radiochemistry. Thus, by genetic engineering techniques, the region of a monoclonal antibody which is responsible for binding the antigen can be produced separately from the rest of the antibody. A whole monoclonal antibody consists of a complex peptide chain arranged in a 'Y' (Figure 22.3). The single vertical section of the Y is a single peptide chain which has a constant structure, common to its class of antibody, and independent of the antibody's antigen-recognizing ability. The two 'arms' of the Y, called F(ab) fragments, consist of double peptide chains and, joined together, are called an F(ab)'2. The individual F(ab) compo-

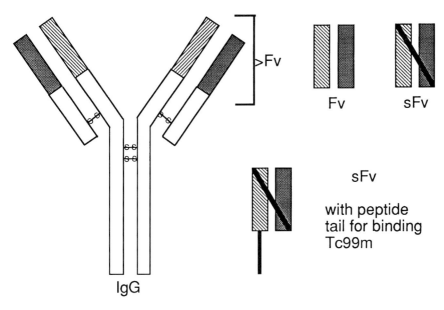

Figure 22.3 Schematic representation of the creation of a monoclonal antibody sFv fragment with an attached peptide tail for technetium-99m binding.

nent consists of a distal variable portion, which recognizes the antigen and a proximal portion, which, like the Fc fragment, is constant. The Fc portion can be cleaved from the whole antibody to leave the F(ab)'2, which, in turn, can be further cleaved to form the F(ab)s. The variable distal portion of the F(ab), called an sFv (single variable chain), can be produced by genetic engineering. Furthermore, it is possible to code in a short tail of a suitable peptide sequence designed to bind a specific radionuclide, and this has now been achieved successfully with Tc-99m (George et al., 1995).

PHYSICS

Whilst nuclear medicine has shifted in recent years from a predominantly physics-based specialty towards one more dependent on chemistry and molecular medicine, physics remains an important foundation and significant advances in physics continue to take place. Single photon computerized emission tomography (SPECT) is an obvious example. It is now agreed universally that in several areas of nuclear medicine SPECT should be used; for example, brain blood flow studies with lipophilic tracers, myocardial perfusion imaging and investigation of unexplained back pain with Tc-99m MDP (Figure 22.4). SPECT based on higher energy radionuclides, such as gallium-67 (Ga-67) and In-111, is also being used increasingly in spite of the much lower photon fluxes than those provided by Tc-99m. It is now generally agreed, for example, that In-111-octreotide imaging is more likely to find elusive primary endocrine tumours if SPECT is employed (Jamar et al., 1995a).

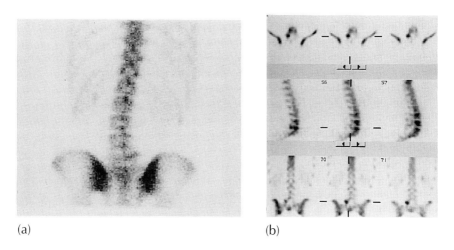

(a)   (b)

Figure 22.4 The comparison of planar (a) and single photon computed tomography (b) images in a patient with low back pain due to a pars interarticularis defect demonstrates improved localization of pathology in the lumbar spine.

Camera design also continues to improve. The analogue-to-digital conversion of the signal from an individual photo-multiplier tube, in contrast to the conversion of the net signal from all of them, improves the spatial resolution and permits the use of thicker crystals. Hence energy resolution and sensitivity, particularly for radionuclides with high-energy emission, is improved and the possibilities of multiple photon acquisition are widened. Another advance is the 511 keV collimator, designed to allow detection of positrons by conventional gamma cameras. This is important with respect to the current interest in co-registration of images which is the superimposition of an image from one agent on that acquired from another. For example, the superimposition of an In-111-labelled leucocyte scan on a Tc-99m-MDP scan (Figure 22.5) aimed at localizing more precisely an infection to soft tissue or bone. Images from one modality may also be superimposed on another; for example, the superimposition of a PET brain image on an MRI image. There is a danger, however, that co-registration will become an infatuation. After all, visual comparison of a radiograph and a bone scan is co-registration in the brain of the viewer and the visual cortex should not

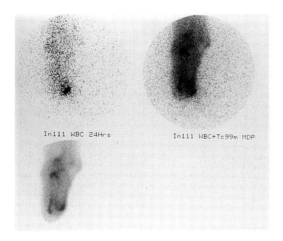

Figure 22.5 Simultaneous dual photon acquisition (Tc-99m-MDP, 140 keV and In-111-granulocyte scan, 174 keV and 247 keV) demonstrating soft tissue infection separate from bone in a young amputee with a stump infection.

be underestimated in its ability to do this. Superimposition of images by simultaneous acquisition, on the other hand, is a fundamental advance since it removes the difficulties related to exact repositioning of the patient. An effective 511 keV collimator and a fully digitized gamma camera could, for example, give high quality simultaneous SPECT and PET images based respectively on Tc-99m isonitrile to define flow and [F-18]FDG to define metabolism, so enabling non-functioning yet viable myocardium to be detected (Burt et al., 1995). Simultaneous imaging across modalities is, however, a greater challenge.

Advances in the design of the PET camera may soon result in a wider availability of PET facilities. A new camera has been developed which, instead of having a full ring of detectors, has opposing segments of detectors (Bailey et al., 1994). By providing slip rings for rotation of the gantry at relatively high speeds, tomographic and whole body imaging can be achieved without the need for a full ring of detectors. This design brings the price of the camera down significantly and into a price range accessible to regional centres that do not possess a cyclotron. Whilst significant distances from a cyclotron unit would prohibit the use of the ultra-short-lived PET tracers, agents such as [F-18]FDG could be transported and used in a variety of clinical settings. This technology will, nevertheless, have to compete with 511 keV collimators.

ADVANCES IN SPECIFIC SYSTEMS

Imaging inflammation continues to be a major challenge for nuclear medicine. Labelled leucocytes, although highly effective, have two major drawbacks. Firstly, there is the requirement for the handling of autologous blood and, secondly, the relative ineffectiveness of labelled leucocytes in chronic inflammation. A further criticism may be their inability to distinguish between sterile and infected inflammation. Several novel tracers have now been explored for imaging inflammation with the aim of labelling granulocytes in vivo or directly targeting the inflammation. These include chemotectic peptides which label granulocyte receptors for mediators of inflammation (cytokines) (Fischman et al., 1994) and anti-granulocyte monoclonal antibodies (Becker et al., 1994). Labelled antibodies (Solanki et al., 1993) and monoclonal antibodies to antigens on the luminal surface of endothelium (Jamar et al., 1995b), on the other hand, directly target the inflammation. Although labelled antibiotics offer the attraction of being able to distinguish between infected and non-infected inflammatory lesions, this is generally not a clinical problem. Labelled antibiotics are therefore unlikely to compete with labelled leucocytes. Furthermore, in fever of unknown origin, the clinical aim is generally to localize pathology rather than to make a specific diagnosis. Labelled antibiotics are therefore unlikely to be as useful as, say, Ga-67. On the other hand, where there are strong clues to suggest an infective cause of a fever, labelled antibiotics may be useful, especially in carefully selected patients. In chronic inflammation, the way forward may well be to image the endothelium, either with monoclonal antibodies to luminal surface antigens or labelled ligands to adhesion molecules upregulated during inflammation (Chapman and Haskard, 1995). Agents which are currently available for imaging chronic inflammation include Ga-67 and human non-specific immunoglobulin (HIG). In-111-

HIG seems to be preferable to Tc-99m-HIG because firstly, localization may take several days and, secondly, In-111, but not Tc-99m, transchelates from the HIG to local extravascular proteins, thereby promoting a higher target-to-background ratio (Claessens et al., 1995).

The main topic of interest in cardiac imaging is the detection of hibernating myocardium. Since thallium-201 (Tl-201) re-injection imaging was introduced, it has become clear that regions which show up as defects immediately following injection at stress and fail to accumulate activity after four hours (fixed defects) are not necessarily infarcts but may be viable myocardium (Dilsizian et al., 1990). The realization of this phenomenon, coupled with the introduction of the non-redistributing Tc-99m-labelled myocardial perfusion agents, has resulted in the introduction of several protocols based on these agents and Tl-201. For the detection of hibernating myocardium, Tl-201 re-injection imaging should be delayed because imaging immediately after injection may still give rise to defects in severely ischaemic myocardium (Dilsizian et al., 1990). The information from such an image is comparable to injection at rest of a Tc-99m agent (such as MIBI). There are, therefore, four types of image available in myocardial perfusion imaging: (1) early imaging after injection under stress (Tl-201 or Tc-99m-MIBI), (2) delayed imaging after injection under stress (Tl-201), (3) early imaging after injection at rest (Tl-201 or Tc-99m-MIBI), and (4) delayed imaging after injection at rest (Tl-201). With Tl-201, all four may differ. It will be exciting to follow developments in 511 keV collimation with respect to the simultaneous imaging of Tc-99m-MIBI and [F-18]FDG (Burt et al., 1995) to see whether digital cameras are able to simultaneously image Tc-99m-MIBI (injected under stress) and Tl-201 (injected four hours earlier at rest).

Exciting developments in oncology include PET of tumours with [C-11]thymidine and [F-18]FDG (Lewis et al., 1994) and the use of Tc-99m-labelled sFv fragments (George et al., 1995). The latter should offer significant advantages over whole antibodies or F(ab) fragments because, firstly, they are less antigenic themselves, and, secondly, as they are smaller, they will penetrate tumours more effectively and will also give a lower background.

CONCLUSIONS

The future of nuclear medicine looks bright but it does depend on keeping ahead of other modalities with respect to novel functional imaging techniques. This will require consolidation of the physics base and also, of great importance, the development of the scientific infrastructure of nuclear medicine, especially molecular biology and chemistry. Such an infrastructure should also ultimately benefit radiology in general and its development should encompass radiology departments as a whole, not just nuclear medicine.

References

Bailey, D.L., Zito, F., Gilardi, M-C. et al. (1994) Performance comparison of a state-of-the-art neuro-SPET scanner and a dedicated neuro-PET scanner. *European Journal of Nuclear Medicine* **21:** 381–387.
Becker, W., Goldenberg, D.M., Wolf, F. (1994) The use of monoclonal antibodies and antibody fragments in the imaging of infectious lesions. *Seminars in Nuclear Medicine* **24:** 142–153.
Burt, R.W., Perkins, O.W., Oppenheim, B.E. et al. (1995) Direct comparison of fluorine-18-FDG SPECT, fluorine-18-FDG PET and rest thallium-201 SPECT for detection of myocardial viability. *Journal of Nuclear Medicine* **36:** 176–179.
Chapman, P.T., Haskard, D.O. (1995) Leukocyte adhesion molecules. *British Medical Bulletin* **51:** 296–311.
Claessens, R.A.M.J., Koenders, E.B., Boerman, O.C. (1995) Dissociation of indium from indium-111-labelled diethylene triamine penta-acetic acid conjugated non-specific polyclonal human immunoglobulin G in inflammatory foci. *European Journal of Nuclear Medicine* **22:** 212–219.
Dilsizian, V., Roco, T.P., Freedman, N.M. et al. (1990) Enhanced detection of ischemic but viable myocardium by the reinjection of thallium after stress-redistribution imaging. *New England Journal of Medicine* **323:** 141–146.
Fischman, A.J., Babich, J.W., Rubin, R.H. (1994) Infection imaging with technetium-99m-labeled chemotactic peptide analogs. *Seminars in Nuclear Medicine* **24:** 154–168.
George, A.J.T., Jamar, F., Tai, M-S. et al. (1995) Radiometal labeling of recombinant proteins by a minimal, genetically engineered chelation site: coordination of technetium-99m by a 26–10 single-chain Fv fusion protein through a carboxyl-terminal cysteinyl peptide. *Proceedings of the National Academy of Sciences* **92:** 8358–8362.
Hawkins, P.N., Myers, M.J., Lavender, J.P., Pepys, M. (1988) Diagnostic radionuclide imaging of amyloid: biological targeting by circulating human serum amyloid P component. *Lancet* **i:** 1413–1418.
Jamar, F., Fiasse, R., Leners, N., Pauwels, S. (1995a) Somatostatin receptor imaging with indium-111-pentetreotide in gastroenteropancreatic tumors: safety, efficacy and impact on patient management. *Journal of Nuclear Medicine* **36:** 542–549.
Jamar, F., Chapman, P.T., Harrison, AA. et al. (1995b) Imaging endothelial cell activation in inflammatory arthritis using an In-111 labeled anti-E-selectin monoclonal F(ab')2. *Radiology* **194:** 843–850.

Lamberts, S.W.J., Bakker, W.H., Reubi, J-C., Krenning, E.P. (1990) Somatostatin-receptor imaging in the localization of endocrine tumors. *New England Journal of Medicine* **323:** 1246–1249.

Lewis, P., Griffin, S., Marsden, P. et al. (1994) Whole body 18F-fluorodeoxyglucose positron emission tomography in preoperative evaluation of lung cancer. *Lancet* **344:** 1265–1266.

Peters, A.M. (1994) The utility of [99mTc]HMPAO-leukocytes for imaging infection. *Seminars in Nuclear Medicine* **24:** 110–127.

Solanki, K.K., Bomanji, J., Siraj, Q., Small, M., Britton, K.E. (1993) Tc-99m 'infection': a new class of radiopharmaceutical for imaging infection. *Journal of Nuclear Medicine* **34:** 119P.

23
Managing Professional Development in Relation to Service Needs

Noelle Skivington

INTRODUCTION

What is professional development? Why must it be managed? What are its benefits for the health service? These are the basic questions to be addressed by this chapter. As a starting point it may be helpful to relate the various buzz words prevalent in human resource circles at the moment, and to review the current climate of organizational development.

There is an incessant drive to maximize output and minimize cost whilst retaining quality standards. As quality measures, particularly in health gain terms, are as yet poorly defined, the responsibility for maintaining and improving this aspect of the service rests firmly with clinical practitioners. This, clearly, includes radiographers whatever their discipline or field of practice. As a fundamental premise, it is vital that professional development enhances the quality of care in a measurable way and does not merely reduce costs.

Within the health care sector it is true that the highest revenue cost is that for staff. The report *Improving Your Image* (Audit Commission, 1995) quantified staff costs as 55% of the revenue costs in a department of radiology. It is also the case that quality of service is determined mainly by staff whether they are using equipment that is twenty years old or only a few days old. These staff need to be able to maximize the use of equipment and to be able to provide a cost-effective, quality service.

CHANGES IN THE WORKPLACE ENVIRONMENT

Organizational Issues

Many studies have looked into skill mix as an obvious area to be included in professional development. The College of Radiographers has issued guidance (Paterson, 1995), as has the Royal College of Radiologists, on various aspects of this. Both bodies, as well as others, recognize that the demand for health care is growing while the resources available for this care are standing still or diminishing. A radical rethink on many skill mix and organizational issues has already occurred within radiology departments. There are also many changes to the processes used in health care delivery generally; patient focused care (Carney, 1995) is one example, as is the move to community based medicine and the moving of diagnostic imaging services into general practitioner (GP) centres. All of these changes are well documented in the report *Getting the Best for Patients* (Welsh Health Planning Forum, 1994) which summarizes current views and makes recommendations for the future. The Calman report on commissioning for cancer services provides a similar, although less embracing, review relative to radiotherapy and oncology services (Expert Advisory Group on Cancer, 1995).

The Individual

All these pressure have changed the cultural environment in which people work. One of the major changes is the need for individuals to develop skills which equip them to adapt to change and to be flexible in their approach to work. The concept of maintaining employability is the current vogue and the job for life philosophy has been questioned (Watkins and Drury, 1994). The need for professional staff to embrace the philosophy of lifelong learning is described by Sir Michael Bett, President of the Institute of Personnel and Development, as 'Achieving qualification is only the start of a journey of lifelong learning ... to uphold the highest standards of professional conduct and to developing themselves ...' (Bett, 1995).

A FRAMEWORK FOR PROFESSIONAL DEVELOPMENT TO MEET SERVICE NEEDS

The framework illustrated in Table 23.1 demonstrates how the links between service needs and professional development can be achieved.

Table 23.1 Overall framework for development

Responsibility	Management tool
Manager	Business plans Corporate strategy ↓ ↓ Departmental strategy ↓
Both	Appraisal system ↓ Individual development/training plans ↓
Individual	Continuing professional development ↓ Personal action plan/career plan

As a departmental manager, the obvious starting point is at the corporate level. However, the individual should have an equal input into the system by starting at, and using, the personal action plan and the continuing professional development programme prior to entering the appraisal part of the framework. Within the framework illustrated, both the manager and the individual, or worker, have their own responsibilities, as well as common or shared responsibilities.

Corporate and Departmental Strategies

A number of terms relative to managing health services are in common use currently. These include 'corporate strategy', 'business plans', and 'departmental strategy' and these tend to be used to identify and to communicate the direction that a particular health service organization is taking. Each one, therefore, is dependent on the local situation and the particular services that have been identified as being required to meet local health care needs.

The Institute of Personnel and Development stress that for learning to be fully beneficial to the organization and its employees 'the organization should have some form of strategic operational plan – and the implications of this plan should be spelt out in terms of the knowledge, skills and concerns of all employees' (Institute of Personnel and Development, 1993).

A departmental strategy will include problem areas identified through such tools as service audit, clinical audit, contract monitoring, complaints received and also potential developments. Most radiology departments will also have received feedback from the Audit Commission report following the national study of radiology departments undertaken recently. Many areas of changing practice in radiography (Nuttall, 1995) have been identified. Nuttall refers to the climate of change brought about by lack of financial resources and advocates the use of audit to ensure that quality standards agreed with purchasers are maintained. She also emphasizes the appropriateness of training and the medico-legal implications. Nuttall's work is worthy of review and enables common themes and one approach to introducing skill mix into a radiology department to be identified.

Appraisal Systems

There are a wide range of systems in use which have been classified broadly as:

1. Peer accountability – that is appraisal carried out by peers and measuring performance, results achieved and resources used. This type of review is frequently carried out in the context of an external framework of professional standards and not in terms of the strategic or tactical needs of an organization. Medical audit is a good example of peer accountability.
2. Peer review and development is also carried out by peers. It, too, assesses performance but focuses on an individual's development. This type of appraisal tends to exclude organizational goals.
3. Performance target setting and review is carried out by a supervisor and measures performance against targets; cascaded downwards. National Health Service (NHS) Individual Performance Review (IPR) is a typical example.
4. Competence assessment and development is appraisal usually carried out by a supervisor, as in performance target setting and review. However, it also includes a training and development aspect. Sometimes a ranking system is used but the aim is to link an individual's aspirations and abilities with the organization's goals. Ricardo Semler, a Brazilian industrialist, reported to a recent

NHS managers conference how such a scheme had worked for his company but with subordinates reviewing managers. His view that this approach had turned his moribund company into a market leader evoked much comment at the conference (*Health Service Journal*, 1995) However, within the health service and within radiology departments the systems that appear to be favoured most commonly are those that focus on superiors performing the appraisal with a mixture of accountability and development being included.

The appraisal, or performance review, therefore, is a key tool in allowing the needs of a department to be shared with the individuals that staff that department. It is, however, a two way process so allowing individuals the opportunity to identify their needs and for the two sets of need to be related together. The Audit Commission report (1995) supports this: 'All staff... should be aware of their contribution to the service being delivered'. Many criticisms are levelled at appraisal systems and, usually, these centre around the type of system; for example systems based on objective setting; the subjective nature of the system in question, and systems which link appraisal to reward. Other questions relate to whether appraisal is a motivator or demotivator of staff, and whether it really is effective. These arguments are well debated by Schemerhorn et al. (1991). All of the points raised have validity but it must be remembered that communication is the key to appraisal and that effective communication is needed all year round, not merely at a yearly appraisal interview.

Individual Development Plans

The outcome of the appraisal should include an agreed development plan for the individual. This plan is distinct from the personal action plan drawn up by the individual which will be referred to later, although in some schemes, especially those based on peer review and development, the plans may well be similar. The agreed development plan will encompass the broad spectrum of training and development needs including professional development areas such as computed tomography, magnetic resonance imaging, ultrasound and patient counselling. Approaches to meeting the identified training and development needs, such as staff counselling, coaching, and mentoring, together with more informal development methods, for example job shadowing, will also be identified. The plan is owned by both the manager and the individual and both need to work to achieve the desired goals. Malloy (1995) indicates that the plan 'combines the flexibility to meet individuals' different starting points and aspirations with the provision of a framework to give guidance and direction. They can also contribute significantly to the establishment of individual development as a key organizational goal.'

Continuing Professional Development

Continuing professional development (CPD) has been defined by Madden and Mitchell (1993) as 'the maintenance and enhancement of the knowledge, expertise and competence of professionals throughout their careers according to a plan formulated with regard to the needs of the professional, the employer, the profession and society'. Furthermore, Madden and Mitchell state that the aim of CPD is 'to provide a profession where members are fully trained and competent to perform the tasks expected of them throughout their careers'.

This may appear to imply that the responsibility for CPD rests with the employer. It should be remembered, however, that CPD is about the individual and his or her responsibility to keep up to date in terms of knowledge and skills. Debate continues as to the exact format of CPD schemes; the measurement of the outcomes of CPD, and the maintenance of competence to practise. Nevertheless, it is clear that CPD will develop as it is of value to all parties involved.

Personal Action Plan/Career Plan

A personal development plan is a clear summary of an individual's learning needs, together with an action plan for meeting them. It is written in the context of the individual's thoughts about his or her job and can be reflective in nature. It may also include comments on other staff members, and on the individual's personal strengths and weaknesses. It may also focus on a change in work or career and it provides the opportunity to reassess the individual's current position and motivation for acquiring new skills and understanding. Due to the reflective nature of the plan there is potential for it to be misused as part of the appraisal process; a factor which highlights the difference between the two development plans referred to in this chapter (the personal action or career plan and the individual development or training plan) and, unhelpfully, these are often given the same title.

VALUE TO THE ORGANIZATION

The benefits to the health service of matching service needs to individual professional development are encompassed in the Investors in People scheme. This is based on the experience of many successful companies which have found that performance is improved by a planned approach to:

> *setting and communicating business goals*
> *developing people to meet these goals*
> *so that*
> *what people can do and are motivated to do*
> *matches what the business needs them to do*
> *(Investors in People, 1992)*

Research has show that the effects of the Investors in People scheme are associated with 'greater formality and explicitness amongst the procedures for marrying the business in general and HR [human resource] issues in particular' (Spilsbury et al., 1994).

The Investors in People approach encompasses all of the management tools identified in the framework outlined earlier but concentrates on the importance of the individual to the success of an organization. The cycle of commitment, planning, action and evaluation stresses that the process is constantly under review and identifies the benefits in organizational terms. These include the now inevitable requirement for improvement in quality, higher levels of staff motivation, increased customer satisfaction and an enhanced reputation for the organization. Indicators include written evidence of commitment such as a training policy statement; business plans reviewed to meet changes which are in turn reflected in the revision of training and development needs; checks on individuals' records to assess this has been carried out, and evidence that training costs have been identified wherever possible. Such an approach seems right, reflecting the changes described earlier.

The need to develop people, particularly radiographers, has been identified clearly (Nuttall, 1995; Paterson, 1995). The process to do so has also been identified in the form of the framework outlined in Table 23.1; and the emphasis on individual responsibility is noted. The role of the employer remains that of supporting, encouraging, motivating and providing guidance for employees. There will be costs involved, not purely in monetary terms but also to create the opportunity and time. These have to be met by organizations, by departments or by individuals. Training and development opportunities remain the one area open to managers to improve working conditions. The key question is where does this rank in the priority list for funding?

Training and development is ranked very highly by Investors in People but is this because professional development compares to the drives to improve effectiveness through altering processes of delivery; processes which have been almost exhausted. There is a feeling that the changes made in the delivery of care have reduced costs to a minimum. Hence, the next and remaining step is to maximize efficiency of the individual and professional development enables this. The one constant question for the managers is what will be the next management tool to be introduced?

CONCLUSION

This chapter has not concentrated on radiography solely but the major issues of skill mix and role development in radiography have been referred to and are well documented elsewhere (Carney, 1995; Nuttall, 1995; Paterson, 1995). However, the principles pertaining to professional development that have been outlined apply to all departments, including radiology. But, for radiographers, they are particularly pertinent due to the already established service pressures to implement professional development (or role development) programmes. All of the management tools referred to need to be used to achieve this. Additionally, the professional body's intention to launch a continuing professional development programme during 1996 will assist managers to manage professional development and to relate this to service needs.

Radiographers, whether as managers or as clinical practitioners must ensure that the profession continues to improve the service they offer to patients and must be able to demonstrate this achievement at a corporate, a departmental and an individual level.

References

Audit Commission (1995) *Improving Your Image*. London: HMSO.
Bett, M. (1995) *Continuing Professional Development*. Institute of Personnel and Development.
Carney, D. (1995) *Current Topics in Radiography No 1*, Chapter 5, London: W.B. Saunders.

Crail, M. (1995) The heretic *Health Service Journal* **108:** 16.

Department of Health (1994) *A Policy Framework for Commissioning Cancer Services – A Consultative Document*. London: Department of Health.

Expert Advisory Group on Cancer (1995) *A Policy Framework for Commissioning Cancer Services*. Department of Health and Welsh Offices.

Institute of Personnel and Development (1993) *The IPD Statement on Continuing Preferential Development: People and Work*. London: IPD.

Investors in People (1992) *Investing in People*. Investors in People.

Madden, C.A., Mitchell, V.A. (1993) *Professions Standard and Competence: A Survey of Continuing Education for the Professions*. Bristol: University of Bristol.

Malloy, J. (1995) Personal development plans. *Management Services* **39(4):** 3–4.

Nuttall, L. (1995) *Current Topics in Radiography No 1*, Vol. 1, Chapter 4. London: W.B. Saunders.

Paterson, A.M. (1995) *Role Development – Towards 2000*. London: The College of Radiographers.

Schermerhorn, Hunt, Osborn (1991) *Managing Organisational Behaviour*, 4th edn. Chichester: J Wiley.

Spilsbury, M., Atkinson, J., Hillage, J., Meager, N. (1994) *Evaluation of Investors in People in England and Wales*. Institute of Manpower Studies.

Watkins, J., Drury, L. (1994) *Positioning for the Unknown – Career Development for Professionals in the 1990s*. Bristol University of Bristol.

Webb, S. (1994) *Indicators Explained*. Investors in People.

Welsh Health Planning Forum (1994) *Getting the Best for Patients*. Welsh Office.

24
Shared Learning: Improved Care and Professional Survival

Hazel Colyer

Since the mid 1970s, the promotion of user centred health and social services has become a privileged plank of the ideology of health care, evidenced particularly in the implementation of the National Health Service and Community Care Act 1990 and the Patient's Charter initiatives of 1992 and 1994. Collaboration and joint working among the different health and social care professions is seen as an essential element for the achievement of this aim.

Consistently during this period, government policy directives and guidelines for good practice have reinforced the view that effective provision of health and welfare services is predicated on accurate needs' assessment and a multidisciplinary, collaborative approach to service delivery. Underpinning this drive for more effective service outcomes through interprofessional collaboration is the widely held view that shared learning opportunities are the key to more effective joint working.

This chapter will attempt to elaborate the nature and strength of the link between shared learning and better services, evaluate the implications for health professions of engaging with shared learning opportunities and the effects on radiography. The term interprofessional shared learning (IPSL), after Barr (1994), is used to obviate the confusion between terms such as interprofessional, multiprofessional and multidisciplinary education or training.

THE CONTEXT

In the past there was a belief that joint working could be implemented by changes to organizational structures; for example the co-terminosity of health and social service boundaries through the establishment of Area Health Authorities in the 1974 National Health Service (NHS) reorganization. This strategy, termed 'organizational fallacy' in a research report (1978) for the 1979 Royal Commission on Health, has been reinvented subsequently as one of the provisions of the NHS and Community Care Act 1990; the Community Care Plan. This joint health authority and social service department plan provides an overview of local community needs and prorities, and is revised annually.

The implementation of a multidisciplinary approach to care and treatment has been difficult and has caused much anxiety to the professions involved, often due to the lack of any shared meanings and value systems. The recognition that shared learning is necessary to support joint working has therefore risen higher on the agenda, resulting in a plethora of initiatives nation-wide ranging from highly focused training workshops around specific topics such as child protection, to fully developed post qualifying master's degree programmes.

Barr's (1994) review of shared learning in the United Kingdom documents current trends through a series of informed impressions which highlight the complexity and differing agendas of shared learning initiatives apart from the classic perspective that 'it (shared learning) encourages interprofessional collaboration which, in turn, provides better integrated care and treatment' (Barr, 1994, p. 8).

However, this ideology is being imposed on services whose organizational structures and professional cultures have become institutionalized since the inception of the NHS in ways which are overtly inimical to joint working. The effect of this may well be that apparently successful shared learning initiatives fail to translate into practice due to structural and professional role constraints.

Additionally, the NHS and Community Care Act 1990 has put in train a series of radical reforms which have shaken the service to its foundation, causing job insecurity and threatening traditional professional practices. The reforms have been construed as a battle for the soul of the NHS against the denizens of general management in the context of post industrial, economically constrained Britain. Such an excess of meaning may well result in manifest resistance and a siege mentality, further hindering interprofessional collaboration.

To date, the radiographic profession has remained largely immune to the IPSL debate, taking a 'what's all this to do with us?' approach. Closely allied to the biomedical, technical model of health care, the diagnostic arm of the profession in particular has viewed the processes of 'hands on' caring as the soft end of medicine. For so long at one stage removed from user centred, needs led care and serving primarily the needs of the medical profession, radiographers, diagnostic and therapeutic, have largely ignored the philosophical and political context of service provision; a situation which cannot continue.

Currently the profession is undertaking a wide ranging review of its role and as it becomes more autonomous within its domain of practice it is growing in self-confidence through this period of change and uncertainty. Professional identity is being defined and refined against allied professions in health and welfare services and this remodelling is being played out in a context which privileges efficiency and effectiveness of service delivery through audit, accountability, consumerism and collaboration. At this time of self-appraisal and self-doubt, the question of engaging with IPSL creates further tensions and professional dilemmas.

INTERPROFESSIONAL SHARED LEARNING

In a uniprofessional learning set who are exploring the theory–practice interface, theory is tested for adequacy by experience in practice and, in its turn, informs and interprets practice. Thus the knowledge base of that profession is developed, extended and transmitted dynamically in a spiralling but professionally circumscribed activity. This process of disseminating the curriculum of expert knowledge within professions is described by Bernstein (1971) as a collection code; individual subjects form a collection of esoteric knowledge which functions additionally to perpetuate societal hierarchies and disempower outsiders.

The value of IPSL and its relative contribution to the achievement of better services for users will depend greatly on the structure and content of the learning experience, and the professional and personal profile of the learning set, although this latter factor is difficult to account for. The acquisition of new knowledge and skills, together with an increase in understanding of other professional roles and the clarification of shared meanings, is the usual expressed desirable outcome of the learning experience.

In IPSL, significant new knowledge may also be synthesized by individual participants through the sharing of experience in the context of theoretical analysis and its application to practice so that the specific learning objectives are enhanced by the diverse perspectives of the group. This occurs as a direct result of extending the parameters of the learning process and interconnecting the different expert knowledges, which Bernstein (1971) dubs an integrated code.

Thus IPSL may open the way to an examination and refining of professional boundaries. However, the benefits to service users are more equivocal. Even if IPSL has proved an enriching educational experience for the individual participant, there is no primary causal link between IPSL and effective joint working which guarantees better services for users except where the development of collaborative strategies for meeting a particular user group's needs is one of the learning objectives of the experience.

In all other cases it can be argued that IPSL is a valuable, 'value added' learning experience which, depending on individual behavioural and organizational characteristics, should enhance professional development, encourage multiprofessional collaboration and further act as a catalyst for the re-examination of professional boundaries. This may ultimately result in restructuring and more genuine interprofessional, user centred care and treatment where knowledge is shared and claims to exclusivity abandoned.

The distinction between multiprofessional and interprofessional collaborative work can be seen to rest on the move from collection code to integrated code; in the former each professional group brings its unique insight and individual contribution to bear on the identified health need, while in the latter there is a sharing of skills and knowledge among the group and the giving up of the exclusive right to undertake particular tasks, an enterprise which is professionally dangerous and may be resisted.

This analysis suggests that the link between IPSL and improved care for users of health services is more tenuous and difficult to sustain than its supporters believe, a claim borne out by two recent studies. Barr (1994) reports that the introduction of structured opportunities for shared learning in qualifying diploma courses for chiropodists, occupational therapists and physiotherapists at Salford was received equivocally: 'while students were well disposed towards shared learning, they felt its objective was not achieved. It had not created the opportunities for interaction which they had expected' (Barr, 1994, p. 26). Interestingly, after sharing in an exploratory, preliminary initiative, radiographers did not take part in this one.

Gill and Laing (1995), in an evaluation of a Diploma in Professional Studies for post registration health visitors, district nurses and occupational health nurses undertaken at Suffolk College in 1991, conclude, 'IPSL is not the panacea for the education of health professionals, and it may even be stated that IPSL promises more than it delivers' (Gill and Laing, 1995, p. 192).

Furthermore, it is not clear that there is any professional advantage to be gained from engaging with IPSL. Indeed, a commitment to it may be professionally unwise if it leads to deconstruction of professional identity and concomitant destabilizing of the radiographic profession.

PROFESSIONS: IDENTITY, BOUNDARIES AND IPSL

What may ultimately be achieved through IPSL is, then, more subtle and linked to the ongoing debates about professions and their relative contributions to health and welfare services. It cannot be denied that the division of labour in the health care industry has shifted perceptibly over the past five years, away from medical consultant dominance and towards general practitioners and the other professions allied to medicine (sic).

This renegotiation of old hierarchies is producing an interesting power struggle fuelled by government policy of transferring resources to community based and primary care services, and also by significant socio-cultural changes which privilege affective individualism, popular democracy, freedom of information and participation in decision-making at all levels.

Carrier and Kendall (1995) summarize the multiplicity of critiques of the traditional professions such as medicine into positive and negative perceptions. In the former, functionalist perspectives, professional practitioners have functionally specific competences which are exercised neutrally, authoritatively and with the best interests of the patient at heart. Professional expertise is gained over a long period of self-regulated education and training and is characterized by discretion in decision-making and beneficence. As Carrier and Kendall say, the members of the profession lay claim to 'sapiental authority' (1995, p. 11).

The negative, structuralist perspectives view professions as located dominantly in a social system of hierarchies, achieving and maintaining their status through social exclusion, thus perpetuating social inequalities. All professions are dominated by the middle classes and the traditional ones, the church, the law and medicine were the province of the upper middle classes and, until recently, have excluded women. This high social position is maintained by 'the exclusive right to offer specific services, a right sustained by the state' (Friedson, 1986, p. 63).

Whatever personal stance is adopted to the above perspectives, they offer valuable and compelling insights into the nature of the radiographic profession. There can be no doubt that radiography, like all health professions, is a powerful interest group; closely aligned to the institutions of the state through registration, acting effectively as gate keeper to services through complex referral systems and with structures which militate against openness and the genuine involvement of lay people. On this analysis, it is not obvious that the profession should share knowledge and skills with other professional groups in order to improve care and treatment for users, if the net result is a diminution in power and influence.

The aforementioned summaries of professions, though illuminating, are largely generic ascriptions which fail to take account of the different levels of professionalism apparent among the health and welfare professions. The changes in the division of labour previously alluded to are being accompanied by increasing claims to professional status by all of the non-medical professional groups, especially in regard to autonomous practice.

It happens that these claims are consonant with government's desire to dilute medical power and are in keeping with socio-cultural shifts in public expectation. It is therefore unsurprising that they are being heard sympathetically. The professional monopoly of doctors is obstructive to a flexible, market led health service and intraprofessional role developments such as radiographic reporting will shortly be well integrated into locally managed health trusts because they offer the degree of flexibility and economic viability which managers want.

This 'doctors against the rest' scenario has three potentially beneficial effects for service users; firstly, the therapy professions such as occupational therapy conceptualize health more broadly and holistically. Recognition of the importance of psychological and sociological influences in the maintenance and restoration of health implicitly locates the user at the centre of the domain of concern and is becoming more influential. Secondly, making common cause against the medical profession increases opportunities for understanding other professional roles. This opening of the communication channels may have the effect of motivating individual professionals to collaborate in the development of joint strategies to optimize care and treatment through IPSL initiatives. Thirdly, the increasing professional confidence attained through reflection on roles and the exercise of political muscle will assist in the definition of strong professional identities and it is only from strength that the possibility of exploring boundaries positively for the benefit of service users can take place.

FUTURE DEVELOPMENT OF THE PROFESSION AND IPSL

The question of whether and how IPSL can create more effective joint working and thus improve services to users is seen to be complex and developmental, dependent on a variety of factors: individual, structural and political.

As discrete events, IPSL initiatives provide excellent opportunities for professional development in the context of shared experience. This may also increase understanding of other professions' roles and responsibilities and break down the barriers of cultural misunderstanding. The new knowledge acquired is enhanced, especially when the subject of the IPSL is generic, such as supervision or management of change, and new skills of collaboration can be developed in highly focused joint training groups, such as child protection or elder abuse. In both of these scenarios, *multi*professional working is facilitated and in the latter it can be achieved.

The will to work *inter*professionally has yet to be demonstrated except in a few, highly selective and well publicized initiatives under the umbrella of 'patient focus'. It involves the sharing of expert knowledge and the giving up of tasks traditionally seen as the exclusive prerogative of a particular profession for the sake of optimizing care and treatment for the user.

Thus, IPSL can be adumbrated as a two stage process, related to whether the endpoint is multiprofessional or interprofessional work. Currently, IPSL initiatives are located mainly in the first (multiprofessional work) stage which, as has been demonstrated previously, may enhance the care given to users through multiple professional perspectives. Progression to the second (interprofessional work) stage is contingent on the success of the first stage and the prevailing sociopolitical context of health service development. It also assumes a willingness on the part of health professions to embrace IPSL at the pre-qualifying level.

This 'second stage' IPSL is a high risk strategy for individuals and professions and will not be entered into lightly by professional bodies who are anxiously securing their position in the new NHS hierarchy. Only when identities are strong and professions secure can there be genuine progress towards the blurring of boundaries.

The health professions can be characterized as resisting interprofessional working because of professional self-aggrandisement, a charge likely to be levelled by politicians and managers as the situation evolves. It can be countered easily in the short term by pointing to service users' right to expect the best possible professional care and treatment which can only be secured through rigorous education and control over practice. In the long term, however, demographic and technological changes will demand new knowledge and ways of working which can only be achieved by engaging with second stage IPSL at the earliest possible opportunity to form new alliances and perhaps new professions.

IPSL IN UNDERGRADUATE EDUCATION

For radiography, intraprofessional changes to work practices are producing a new generation of autonomous practitioners able to offer a more comprehensive and efficient service to their users and benefiting from IPSL initiatives in that endeavour. Genuine interprofessional work would refine further the shape of the profession in the 21st century, moving users even closer to the heart of the decision-making process in ways which are not clear or comfortable.

The first steps towards becoming an integral and defining part of this process must be to integrate IPSL into the pre-qualifying, undergraduate education programmes of radiographers to prepare them for willing collaboration with other professions, whilst building a strong professional identity. A common base and the chance to develop shared values would abolish the contradictions at the heart of many current 'first stage' IPSL initiatives: those of professionalism versus pluralism, status versus parity of esteem and professionally focused versus user focused services.

An example of undergraduate IPSL has been implemented in the Faculty of Health Sciences in Linkoping, Sweden. Using problem based learning strategies, a ten week joint introductory programme for six different professional groups was devised and followed up with intermittent IPSL sessions interspersed in the professional education programmes of each individual group (Areskog, 1995). Evaluation of the initiative has been encouraging and a number of other projects have been developed subsequently. Undergraduate IPSL has been received more positively by students than lecturers and Areskog also reflects that professional bodies have been sceptical, probably for reasons previously elaborated.

CONCLUSION

There is now beginning to be in the United Kingdom an irresistible pressure for health services which are needs led and sensitively exercised in favour of users rather than providers. This chapter has demonstrated that quality shared learning opportunities are a key feature in achieving this end because they encourage and enhance multiprofessional working, but in ways which are complex and professionally risky.

The realization of interprofessional working would further the process of instituting user centred health services and the current climate of interprofessional jostling for position amidst political uncertainty is opening up the possibility of genuine, 'second stage' IPSL, entered into from a position of strengthening professional autonomy, though the time is not yet opportune.

As professional identities grow stronger, the willingness to collaborate and negotiate boundaries will increase and professions will begin to bow to public expectation. The shape of health professions in the 21st century is unclear presently. If radiography is to play a full part in determining that shape, then a proactive stance towards developing IPSL opportunities for radiography undergraduates with other professional groups is an essential step in the process.

References

Areskog, N-H. (1995) Multiprofessional Education at the Undergraduate Level. In Soothill, K., Mackay, L., Webb, C. (eds) *Interprofessional Relations in Health Care*, pp. 125–139. London: Edward Arnold.
Barr, H. (1994) *Perspectives on Shared Learning*. London: CAIPE.
Bernstein, B. (1971) *Redefining Professional Boundaries through Education*. London: Routledge.
Carrier, J., Kendall, I. (1995) Professionalism and interprofessionalism in health and community care: some theoretical issues. In Owens, P., Carrier, J. and Horder, J. (eds) *Interprofessional Issues in Community and Primary Health Care*, pp. 9–36. Basingstoke: Macmillan.

Department of Health (1990) National Health Service and Community Care Act. London: HMSO.

Friedson, E. (1986) *Professional Powers: a Study of the Institutionalisation of Formal Knowledge*. Chicago: University of Chicago Press.

Gill, J., Laing, J. (1995) Interprofessional shared learning: a curriculum for collaboration. In Soothill, K. et al. (eds) *Interprofessional Relations in Health Care*, pp. 172–193. London: Edward Arnold.

Royal Commission on the NHS (1979) *Research Paper No 1*. London: HMSO.

25
Computed Radiography – Influences on Sensitivity and Latitude

Sue Farmer

COMPUTED RADIOGRAPHY

Conventional film/screen radiography (FSR) has provided a means of image capture, display and storage for most of this century. Current film/screen combinations aim to provide good contrast, whilst allowing reasonable latitude to cope with exposure variations and thus make use of the dynamic range of the region of interest (Balter, 1990; Jennings et al., 1992). However, despite advances in the manufacture of such systems, they remain relatively inflexible to exposure variation and cannot offer post-processing facilities. Film's limited dynamic range may restrict the flexibility of such conventional systems.

The introduction of computed radiography (CR) as a means of recording the medical image has brought with it a whole new concept of image production. Such image production steers away from many of the traditional photographic theories and practicalities, yet aims to achieve equivalent high quality diagnostic images. Despite CR being a relative newcomer to the world of medical imaging, its development has been documented since the early 1980s (Cocklin and Lams, 1984; Wijnands, 1984) and its capabilities continue to be explored (Newton, 1995). CR is taking shape as an accepted member of the portfolio of imaging techniques in use across the spectrum of medical imaging.

Whilst the end product for both FSR and CR systems will invariably be a hard-copy radiographic image (except where a CR system is linked to a Picture Archiving and Communication System or PACS), CR's performance varies from that of FSR, particularly when image enhancement is an option. Such enhancement may be closely linked to the sensitivity and the latitude of the CR image (Murphey et al., 1992).

SENSITIVITY AND LATITUDE

What have sensitivity and latitude values traditionally indicated about the images produced using existing film/screen combinations? Both have long been important issues, particularly when dose to the patient and image quality are to be weighed up. Information about the specific behaviour of a particular film type to exposure, where log relative exposure is plotted against density (Curry et al., 1990), can be assessed from the relevant characteristic curve. Increasing efficiency of screen output in relation to the exposure received has contributed to the overall efficiency of FSR, as well as to dose reduction possibilities. However, Greene and Oestmann (1992), amongst others, reflected on the characteristic curve of film and noted its intolerance towards wide variations in exposure. Whilst some ability to cope with exposure variation does occur variously between film types, there are still limits as to how far a radiographic exposure can be altered, before noticeable increase or loss of density renders a radiograph of little diagnostic use.

So, after continuing efforts to produce a film/screen combination with adaptable latitude, yet acceptable contrast and adequate sensitivity, how has the advent of digital imaging techniques such as CR affected the capabilities and response of an imaging system?

CR IMAGE PRODUCTION

In consideration of CR, its principles of operation clearly play an important role in image acquisition. A CR system is based on the use of image storage phosphor, or imaging, plates (IPs), in place of conventional film/screen cassettes. Such plates contain a photostimulable phosphor layer of, for example, fine-grain europium doped barium fluorohalide crystals – BaFBr:Eu (Workman and Cowen, 1995). Following exposure, the latent image held within the phosphor layer is released when the plate is inserted into an image reader and scanned by a helium-neon laser. In this process, light is emitted as photo-luminescence which will then be photo-multiplied and converted to an analogue electrical signal. In turn this will eventually be converted into a digital signal (Hillen et al., 1987).

Following the laser scan, some systems may display the full scan on a work station, together with visual representation of a selection of parameters. These may be selected by the operator and applied to the image, so that the required appearance is achieved – the image may then be printed onto hard copy film. Other systems will obtain a sampling 'pre-read' scan before analogue to digital conversion, whereby data is extracted from the full scan and algorithms are used to assess the useful range of data (signal range). Schaefer and Prokop (1993) found, in this case, that once the relevant exposure range has been detected, only this range would be used in any later image processing or display. The extracted data will then be applied to pre-set parameters which are in histogram form. The format of the histogram is largely determined by the selection of a pre-set code, such as those used with the Fuji AC-1 Plus CR system (Carty and Barrow, 1993), which are designed and selected according to the region of the body under examination. It is such codes which determine the processing parameters for the image and thereby the appearance of the final image. Heath (1995) described how the image processor or reader requires information on the examination type and region of interest prior to the laser scan, in order for the appropriate histogram to be used. Each designated region of interest, for example chest, abdomen or cervical spine, will have a particular associated set of processing parameters for the area under examination.

A pre-read scan may also give important information about the reader sensitivity (S) and latitude (L) values of the image data, which will determine the reading conditions for the full scan. In addition to determining these values, the extent of the irradiated field on the IP may be established, so that irrelevant data, such as that from secondary radiation incident on the IP but outside the collimation, is not included in the histogram analysis which occurs during a pre-read scan. In a review of CR systems incorporating Fuji's 7000 series CR readers, Cowen et al. (1993) commented on the use of an algorithm known as 'pattern recognizer for iris of exposure field' (PRIEF). Such an algorithm may be used to identify the presence of collimation around the area of examination. If collimation has been used, histogram analysis will occur within the collimation; where collimation is absent, the whole field of the IP is analysed. From the histogram analysis the position of the useful data and its proportion within the dynamic range can be assessed. In the same study, Cowen et al. included the example of the possible exclusion of data related to exposure of the IP plate where there has been direct exposure incident on it, that is where the beam has not been attenuated. Additionally, Murphey et al. (1992) described histogram analysis over a 'selected portion' of the image from the IP, so that the diagnostically useful exposure range can be used to best advantage.

Once the S and L parameters have been determined and applied to the final scan, according to the required appearance of the final image, the opportunity for image optimization becomes available. Where histogram analysis 'mismatch' occurs, S and L levels may be adjusted so that the best possible data match is possible; following this, A–D conversion can occur. The S and L values will be printed on the hard-copy final image.

HISTOGRAM ANALYSIS

In consideration of the useful data available from CR images, Schaefer and Prokop (1993) describe the use of specifically designed algorithms as a means of analysing the histogram of the relevant diagnostic exposure range. Such histogram analysis plays a crucial role in the ability of the CR system to offer a wide dynamic range. Following the initial sampling scan, where used, enhancement parameters may be applied which will largely be guided by the required image appearance. The pre-read scan may also play a part in assessment of the overall range of pixel values, as well as the area of exposure.

The selection of processing parameters plays a critical role in the production of the final optimized image, where selective enhancement of the required tissues will occur (McNitt-Gray et al., 1993). When such data are displayed in histogram format, whereby frequency of pixel occurrence is plotted against pixel value, an indication may be gained of the latitude and sensitivity of the image (Figure 25.1). This will vary according to the pre-set parameters and associated histogram, and the required image appearance of each part of the body.

If the histogram analysis representing the actual data received from the IP closely resembles that of the pre-set parameter histogram and the S and L levels are appropriate for the required pixel range and value, little alteration or optimization of the image may be required. However, considering the many variables in imaging conditions for each patient and examination, this is unlikely to be the case.

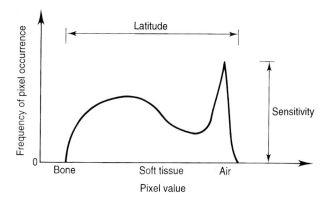

Figure 25.1 Example of a typical CR histogram.

HISTOGRAM APPLICATION

Where histogram analysis does identify a mismatch between the S and L values of the data from the IP and that of the pre-selected parameters, image optimization may be necessary in order to produce the appropriate image appearance.

The histogram shape may be affected by the area of collimation and the range of pixel values contained within. This may affect the use of the previously mentioned PRIEF algorithm. In addition, factors such as parameter and exposure factor selection may affect the shape of the histogram, and in turn the S and L values demonstrated, such that they differ from the pre-set parameters.

Parameter Selection

Should the incorrect processing parameters be selected, the associated histogram may have a different pixel value emphasis to that of the area being examined. This may have critical importance where, for example, a soft tissue abdomen code is applied to a barium sulphate procedure (for example a barium follow-through examination). The barium examination image would by nature include a high frequency of pixel value occurrence corresponding to high density barium sulphate. This would therefore have a different histogram analysis in comparison to that of plain abdominal parameters, where soft tissue values are likely to be emphasized.

Murphey et al. (1991) commented on the difficulties of establishing parameters which provide appropriate enhancement for examinations such as contrast agent procedures. Such substances, for example barium sulphate, will affect the histogram analysis, such that contrast examination images may be degraded. They suggested that enhancement parameters could be modified with available software.

Schaefer and Prokop (1993) commented on the use of correct selection of an appropriate set of parameters in chest radiography and on the reliability of such an algorithm, except where the histogram form is caused to be 'grossly altered'. They suggested a situation where one lung may be completely opacified as an example. Whilst an appropriate histogram analysis may have been carried out, where there is major disparity in the histogram shape, problems with image enhancement may be unavoidable.

Exposure Factor Selection

Incorrect selection of exposure factors has invariably had an important effect on the production of an acceptable diagnostic image with FSR. Whilst some degree of latitude may allow exposure variation, invariably the image produced by FSR is likely to lose diagnostic value, due to under or over exposure. Murphey et al. (1991) suggested that the traditional relationships between exposure factors are altered with CR, as the system is able to automatically alter the S and L values according to the image data. A disparity in the pre-set histogram's frequency of occurrence of pixels may be caused, for example in the case of over-exposure, so that the shape of the histogram is changed and the sensitivity of the image may be altered. The S level, displayed on the final hard copy image, will vary according to this alteration of the histogram. Cowen et al. (1993) suggested caution in the interpretation and use of S numbers or values as an indication of appropriate exposure selection. They suggested using such information as an approximate guide only, whilst taking technical and patient-related conditions into account.

In considering the latitude of CR systems, Newton (1995) experimented by using a step wedge and varying exposure levels, from 2 to 200 mAs, where density values obtained from final hard copy CR images were plotted against exposure. Despite the varying exposure, only narrow ranges of density for the mean highest and lowest density values were obtained, for example excessively high density was not shown, despite selection of relatively high exposure values (200 mAs).

This would suggest that despite wide exposure variation, the latitude of the CR system is still able to produce films of diagnostic quality and this certainly appears to be one of the attractions of CR (Fuhrman et al., 1990). However, this must be seen in tandem with the importance of correct selection of processing parameters.

Heath (1995) discussed exposure factor selection as posed by Murphey et al. (1992) who looked at the possibilities of exposure reduction. They commented on the effects of signal-to-noise ratio and how for example lowered exposure factors caused noise levels in CR examinations of the hip, pelvis and spine. Such noise levels were sufficiently disruptive to cause images to lose diagnostic accuracy at fifty per cent of the original exposure value. However, they also suggested that extremity radiography might be suitable for considerable exposure reductions (up to 50%), whilst maintaining image quality. Cowen et al. (1993) suggested that exposure reduction is limited by image quality demands, including the signal-to-noise ratio, and this remains an issue for further debate and analysis. How far can exposure factors be reduced before it is not possible to produce a diagnostically useful image?

Secondary Radiation

Another factor to be considered in exposure factor selection, as Carty and Barrow (1993) commented in their evaluation of a CR system, is that the image reader cannot distinguish between primary and secondary radiation. A high proportion of secondary (scatter) radiation incident on the IP may cause a significant alteration in histogram shape. Where possible, secondary radiation needs to be kept to a minimum and this may be accomplished by appropriate kV selection.

CONCLUSION

Whilst the FSR combinations in current use provide efficient imaging media, developments such as CR have provided a means of furthering the capabilities of diagnostic imaging processes. As Heath (1995) and others have commented, digital radiography, including CR, will play an important role in the future and suggests that many departments may eventually convert partially or completely to digital systems. Discussion continues on the possibilities of dose reduction to the patient, balanced against image quality. CR plays a major role in the possibilities of image enhancement in general radiography. Where close matching of histogram analysis between the pre-set parameters and those of the image data occurs, high quality images may be assured. Systems incorporating on-screen application of parameters before a hard copy image is produced allow further flexibility. Attention to correct processing parameter selection; reduction of secondary radiation; and appropriate exposure factor selection may help to contribute to appropriate maintenance of sensitivity and latitude values.

References

Balter, S. (1990) An overview of digital radiography. In *Digital Imaging in Diagnostic Radiology*, pp. 81–106. New York: Churchill Livingstone.

Carty, H., Barrow, S. (1993) *Evaluation Report: Fuji AC-1 Plus Computed Radiography System*. London: Medical Devices Directorate, MDD/93/29.

Cocklin, M.L., Lams, P.M. (1984) Digital radiology of the chest. In *Digital Radiology – Physical and Clinical Aspects*, pp. 157–168. London: The Hospital Physicists' Association.

Cowen, A.R., Workman, A., Price, J.S. (1993) Physical aspects of photostimulable phosphor computed radiography. *British Journal of Radiology* **66(784):** 332–345.

Curry, T.S., Dowdey, J.E., Murry, R.C. (1990) *Christensen's Physics of Diagnostic Radiology*, 4th edn. Philadelphia: Lea and Febiger.

Fuhrman, C.R., Gur, D., Schaetzing, R. (1990) High-resolution digital imaging with storage phosphors. *Journal of Thoracic Imaging* **5(1):** 21–30.

Greene, R.E., Oestmann, J.-W. (1992) *Computed Digital Radiography in Clinical Practice*, 1st edn. New York: Thieme Medical Publishers.

Heath, M. (1995) Digital radiography – the future? *Radiography* **1(1):** 49–60.

Hillen, W., Schiebel, U., Zaengel, T. (1987) Imaging performance of a digital storage phosphor system. *Medical Physics* **14(5):** 744–751.

Jennings, P., Padley, S.P.G., Hansell, D.M. (1992) Portable chest radiography in intensive care: a comparison of computed and conventional radiography. *British Journal of Radiology* **65(778):** 852–856.

McNitt-Gray, M.F., Taira, R.K., Johnson, S.L., Razavi, M. (1993) An automatic method for enhancing the display of different tissue densities in digital chest radiographs. *Journal of Digital Imaging* **6(2):** 95–104.

Murphey, M.D., Huang, H.K.B., Siegel, E.L. et al. (1991) Clinical experience in the use of photostimulable phosphor radiographic systems. *Investigative Radiology* **26(6):** 591–597.

Murphey, M.D., Quale, J.L., Martin, N.L. et al. (1992) Computed radiography in musculoskeletal imaging: state of the art. *American Journal of Radiology* **158:** 19–27.

Newton, S. (1995) Conventional radiography versus computed radiography: a study of image quality. *Radiography Today* **61(694):** 21–24.

Schaefer, C.M., Prokop, M. (1993) Storage phosphor radiography of the chest. *Radiology* **186(2):** 314–315.

Wijnands, D.T. (1984) Some aspects of digital imaging and archiving in radiological departments. In *Digital Radiology – Physical and Clinical Aspects*, pp. 169–180. London: The Hospital Physicists' Association.

Workman, A., Cowen, A.R. (1995) Improved image quality utilizing dual plate computed radiography. *British Journal of Radiology* **68(806):** 182–188.

AUTHOR'S REPLY TO EDITOR'S COMMENT ON CHAPTER 15

The clinical, legal and ethical issues involved in medico-legal cases of cerebral palsy are extremely complex and cannot be detailed in the remit given of 2500 words. This chapter could only outline the role of MRI and identify important topics of discussion. It was not possible to evaluate the rights and wrongs of the legal system or the ethics of medico-legal medicine – indeed that was not the purpose of the text. As the principal clinical team involved in these cases we felt, therefore, that some of Paterson's comments were harsh and reflected badly on our practice. We question whether this is an appropriate forum for her to air her personal views and believe that it is important for us to reply.

The fact that parents must pursue legal action to obtain the trust is undesirable, but in the absence of a better system we must work within the present medico-legal framework. Undoubtedly the process of litigation is flawed and parents should not have to go through it in order to gain compensation. Alternative methods have been investigated, but all have their drawbacks. For example a no-fault compensation scheme is more equitable but the cost is high and cannot be fully supported by most healthcare systems. In addition, although a no-fault scheme enables doctors to speak more freely to their patients after a medical accident the present system ensures that, because of the threat of litigation, doctors practice as carefully as possible. In a no-fault scheme this 'policing method' does not operate with a potential for a decrease in clinical standards. Although these cases are tackled via the legal system, it must not be forgotten that the major process of law is to determine the *truth* and to see that justice is done. Many cases come to law long after the event and therefore objective data may be lacking. Surely it is better that settlements are based on the best objective data and, in imaging terms, this means MRI.

Paterson implies that the opinion of the radiologist has influenced others on the clinical team. We wish to state that the radiologist, radiographers and anaesthetist all find the practice acceptable and have not been coerced in any way. We believe that these studies are ethically justifiable on the following grounds:

- Our practice ensures that minimal harm comes to the child. In our hospital the administration of a general anaesthetic is deemed safer than sedation for the reasons outlined in the text. In fact, recent guidelines issued by the Royal College of Radiologists state that if a Radiologist is required to sedate and monitor a patient in addition to overseeing the investigation, there is a conflict of interest which is potentially detrimental to safety. By anaesthetizing these children, therefore, we are not only conforming to the College guidelines, but also ensuring the safest possible outcome.
- As in our hospital it is standard medical practice to anaesthetize children for an MRI examination and MRI has been shown to be the best technique for evaluating brain injury, ethics approval is not required.
- By assisting the legal case, which may ultimately lead to a financial award, we increase the autonomy of the child by providing him or her the freedom to live more independently. Before MRI, it was notoriously difficult to prove the causal link between asphyxia and cerebral palsy. Therefore these children received no compensation and little explanation of events. MRI has facilitated both, which is surely to their benefit.
- We are respecting the wishes of parents and children by performing these examinations – to do otherwise would, in our opinion, be ethically wrong.
- By examining children with cerebral palsy we hope to increase our understanding of the aetiology of the disorder which may ultimately benefit society.

Although an MRI scan is organized by the legal team, instruction to do so is given by the medical experts involved in the case and the anaesthetist takes full clinical responsibility. He or she meets with the parents before the examination to explain the risks of the procedure and informed consent is always obtained. After the scan the radiologist spends considerable time discussing the results of the investigation and most parents state that they find the procedure a very helpful and positive experience. In fact the results of the study often clarify some important issues and alleviate worries they may have.

It is obvious that all health-care workers should maintain their skills and remain abreast of developments but mistakes are made even by the most professional staff. A negligent act does not necessarily mean incompetence and how to differentiate these terms is a very difficult issue. This chapter was intended to open the debate on the role of

MRI in medico-legal medicine and it is evident that the arguments will continue for some time.

Catherine Westbrook
Education and Research Co-ordinator, University of Oxford
Philip Anslow
Consultant Neuro-Radiologist, Radcliffe Infirmary, Oxford
John Stevens
Consultant Paediatric Anaesthetist, John Radcliffe Hospital, Oxford

Index

511 keV imaging 175, 177, 180, 184, 189
   collimators 264, 265, 267
Abortion (Amendment) Act, 1990: 64, 65
Accident and Emergency departments
   Patient's Charter 99
   'red dot' systems 79
Adolescent idiopathic scoliosis (AIS) see
   Scoliosis
Annual occupational dose limits 168–169,
   169–170, 171
Anonymity, data collection 90
Anxiety, cancer patients 105, 115
   during radiotherapy treatment 106
   perception of side-effects 107
   reducing through intervention 108, 109
   see also Macmillan radiographer, the
Appraisal systems 273–274
Artefacts, breast MRI 211–212, 213, 214
Association for Improvements in
   Maternity Services (AIMS) 66–67
Audit Commission 136–137, 274
Autonomy 7, 8, 87, 88

B-mode ultrasound 241, 242
Back pain, unexplained, Tc-99m MDP
   imaging 264
Back shape, ultrasound assessment 241,
   245–246
   relationship to lower limb torsion 246
Basal nuclei 152
Biochemistry, breast cancer screening
   18–19
Birth asphyxia, MRI 148, 149, 150, 151,
   152
   case reports 156–160
Bismuth germanate (BGO) scintillator
   178

Board of Registration of Medical
   Auxiliaries (BRMA) 2, 3
Brachytherapy 38
Brain
   choroid plexus cysts (ultrasound) 61, 62
   FDG gamma camera imaging 188–189
   functional MRI (fMRI) 54
   medico-legal work in cerebral palsy,
   MRI 148–162
Breast cancer
   difficult cases 222
   multifocal/multicentric disease 215
   occult disease 220–222
   recurrence versus scar, MRI 216–218,
   219
   screening 14–24
      biochemistry 18–19
      comment 22–24
      mammography 14–18, 22, 24,
      197–198, 205, 215–216
      suitability of disease 19–21
   treatment 23–24, 43, 196–197, 215,
   225
   patient anxiety 105, 106
   see also Magnetic resonance imaging
   (MRI): breast
Breast conservation surgery 197, 212,
   215, 225
Breast localization coil, MRI 224–225
Breath-holding, MRI 229–230
British Medical Association (BMA) 2, 3
Budgets, departmental 131

Calman report 36, 109, 271
Cancer
   breast see Breast cancer
   FDG gamma camera imaging 188, 192

Cancer (*Continued*)
  incidence and radiation exposure 167, 168, 169
  the Macmillan radiographer 115–121
  patient anxiety 105, 106, 107, 108, 109, 115
  prostate, dose escalation study 42
Cancer Relief Macmillan Fund 116
Cancerlink report 116
Carbon-11 (C-11) labelling 262
Cardiac imaging
  gamma camera 181, 184
    left ventricular myocardium 186–188, 189, 267
  MRI, coronary artery 228–234
Cardiac motion artefacts, MRI 212
Castejon study 117
Cell labelling 261–262, 265, 266
Cerebral palsy, MRI 148–162
Choroid plexus cysts 61, 62
Code of Professional Conduct 8, 10, 145, 162
  ethical issues 83–84, 93
    confidentiality 89–90
    conscientous objection, termination of pregnancy 64–65
  image reporting 81
Codes of Practice, ionizing radiation regulations 165, 166
'Coincidence detection' (PET) 178, 190
Collection code, ISPL 281, 282
Collective dose 28, 29, 168
College of Radiographers 7, 8, 9, 10, 31, 136
  ethical code 93
  misconduct 10, 11
  skill mix, guidance on 271
  see also Code of Professional Conduct
Collimators
  dynamic, treatment beams 41, 43
  gamma cameras 177, 179–180
    511 keV 264, 265, 267
    fan beam 182
  Multileaf Collimator (MLC) system, linear accelerators 38, 39, 40, 41, 42, 43, 44
Communication 274
  image reporting 77
  radiotherapy departments 106, 107–110, 116
  see also Macmillan radiographer, the

Communication (*Continued*)
  results of ultrasound scan 63–64
  in risk management 146
Community hospitals, imaging 254
  equipment 124
  reporting and results 251
  teleradiology 254, 255, 256
Competence assessment and development 273–274
Complaints 134
  malpractice claims, USA 143–144
  Patient's Charter 142
Computed radiography 289–290
  image production 290–292
  sensitivity and latitude 290
  treatment verification radiographs 38
  see also Teleradiology in primary care
Computed tomography (CT)
  asphyxia 160
  breast cancer 199
  radiological protection 26–33
    dose reduction, general trends 30–31
    growth in CT practice 27–30
    quality criteria 31–32
  scoliosis 240
  simulators 37
Confidentiality, data collection 89–90
Conformal radiotherapy 41, 42
Continuing professional development 11, 45, 145, 275
Contracting for quality of care 130–137
  areas of quality 133–134
  audit commission report on radiology 136–137
  external contracting 132–133
  internal contracting 131–132
  monitoring, involvement of professional in 134–136
Contrast agents
  gadolinium in MRI 54, 200–201, 208–209
  histogram analysis, computed radiography 294
Coronary artery imaging, MRI 228–234
Corporate strategies, professional development 272–273
Costs
  breast MRI 224
  digitizing equipment 253, 256
  mammography 14–15

Costs (*Continued*)
  negotiating and maintaining standards 134
  PET scanner 185–186
  professional development 276–277
  of risk 142–143
  in service level agreements 132
  staff 270
  to GP fundholders, purchase of services 132–133
Council for the Professions Supplementary to Medicine (CPSM), Radiographers' Board 7, 8, 11, 81, 93
Counselling cancer patients 118
Crown Immunity, loss of 141, 144
Crystals
  gamma cameras 176–177
  PET scanners 178, 186

Declaration of Helsinki 87
Departmental strategies, professional development 273
Deterministic effects of radiation 165
DICOM standard 253
Doppler ultrasound, breast 199
Dose
  from x-ray examinations 27
  reduction, computed radiography 295
  sharing, staff 168
  see also Computed tomography (CT): radiological protection
Dual-headed gamma cameras 177, 181, 183–184
Dynamic collimation, treatment beams 41, 43
Dynamic scanning
  MRI breast scanning 204, 206–212, 215
    contrast enhanced 209–210
  PET scanners 181

Echo planar imaging (EPI) 51, 229
Education
  centres, teaching risk management 145–146
  ISPL in undergraduate 286–287
  see also Training
Electronic portal imaging (EPI) system 40, 41

Enhanced breast MRI 200–205
  dynamic imaging 209–210
  enhancement patterns 203, 204
  why tumours enhance 201–203
Errors and omissions in reporting 76
Ethical
  dilemmas in radiographic research 83–94
    comment 92–94
    guiding moral principles 85–86
    Kantianism 85
    research stages and moral issues 86–91
    utilitarianism 84–85
  issues
    MRI in cerebral palsy medico-legal work 155
    ultrasound, fetal anomalies 58–67
Ethics committees 92, 162
European Commission (EC), radiation dose limits 170, 171
Expert Advisory Group on Cancer (EAGC) 111
Eye, lens of, dose equivalent limits 165–166, 168, 170, 171

Fast gradient echo sequences 210
Fast spin echo (FSE) imaging sequences 53
Fat saturation, breast MRI 211
FDG (fluorine-18 fluorodeoxyglucose) 174, 179, 181, 190, 191–192, 266
  imaging left ventricular myocardium 186–188
Femoral anteversion, ultrasound evaluation 241, 244, 246
Film/screen radiography (FSR) 289, 294, 295
Flow quantification, MRI of coronary arteries 234
Forrest Report 14, 15, 16–18, 22, 23
Functional MRI (fMRI) 54, 56

Gamma camera imaging, positron emitting isotopes 174–192
  clinical FDG gamma camera imaging 191
  left ventricular myocardium 186–188
  oncology and brain 188–189
  comparative imaging 179

Gamma camera imaging, positron
  emitting isotopes (*Continued*)
  count rate capability 182–183
  detection sensitivity 180–182
  discussion 189–191
  gamma camera and PET scanner
    technology 175–178
  gamma ray collimation, effect on
    performance 179–180
  multiple isotope studies 183
  quantitative imaging 183
  radiopharmaceuticals 179
  recent developments in technology
    183–184
  spatial resolution 182
General anaesthesia (GA), MRI of
  children 155, 161
General practices, imaging services
  125–126, 126–127, 129
  fundholding (GPFH) 124, 125, 127–128,
    129, 131, 132, 257
  purchasing services 127–128
  see also Teleradiology in primary care
Gradients, MRI 49

Haematoma, breast, MRI 222, 223
Hand imager, small bore dedicated MRI
  50
Helical (spiral) scanning (CT) 30, 32–33
Hibernating myocardium 186, 187, 189,
  267
High field MRI 50
High resolution MRI 53
  3D volume imaging 210, 211
High sensitivity 3D imaging, PET 185
High speed imaging, MRI 51, 53
Histograms, computed radiography 291,
  292
  analysis 292
  application 293–295
    exposure factor selection 294–295
    parameter selection 293–294
Hybrid gamma camera/PET scanners 190
Hypervascularity, tumour 202–203
Hypoxia–ischaemia (HI) 150–151,
  152–153
  case reports 156–160

I-123 labelled serum amyloid P
  component (I-123–SAP) 261

Image quality, teleradiology 255–256
Image reporting 2, 8, 70–81, 284
  comment 78–81
  concepts in report 72–74
  definition of report 70–71
  descriptive/interpretive elements 71, 72
  'findings', meaning of 72
  model for 74–75, 80
  in outreach services 127
  purpose of report 77–78
  quality of report 75–76, 80
Incidence of cancer, radiation exposure
  167, 168, 169
Independent Inquiry into Obstetric
  Ultrasound Procedures 67
Indium-111
  In-111–HIG 266–267
  In-111–labelled octreotide imaging 260,
    264
Individual development plans 274–275
Inflammatory lesions, nuclear medicine
  imaging 266–267
Information 102
  radiotherapy patients 108, 110,
    117–118
  lack of 106, 107
  withholding information 62, 161
Informed consent
  fetal ultrasound 60–61
  research programmes 87–88, 89
Integrated code, ISPL 282
Integrated Services Digital Network
  (ISDN) link 252–253
International Commission on Radiological
  Protection (IRCP) 168, 171
  dose limits 164, 165
    annual 169
    pregnancy 166, 169
Interprofessional shared learning (IPSL)
  see Shared learning
Investors in People scheme 276, 277
Ionising Radiation Regulations 26, 165,
  166

Life Span Study Group, Japan 167

Macmillan radiographer, the 115–121
  development of role 119–120
  education and research 118–119
  the way forward 120–121

Magnetic resonance imaging (MRI) 29–30
  breast 196–226
    applications 212–224
    development of 199–205
    the future 225–226
    guided biopsy 224–225
    limitations 222–223
    technique 205–212
  coronary artery 228–234
    assessment of coronary stenosis 233–234
    comparison with x-ray coronary angiography 230–233
    future directions 234
    technqiues 229–230
  future developments 48–56
    hardware 49–51
    imaging 51–54
    machines of future 55
    open architecture 37, 51, 52
  medico-legal work in cerebral palsy 148–162
    case reports 156–160
    comment 161–162
    ethical issues 155
    the legal process 149
    role of MRI 151–154
Malpractice claims 143–144
  see also Misconduct; Negligence
Mammography 14–18, 22, 24, 197–198, 205, 215–216
  cancer triggering 17–18
  dense breasts 198, 216
Mastectomy 197, 220
Maximum permissable dose (MPD) 165, 166
Merrett Health Risk Management Ltd manual 142
Meta-iodobenzylguanidine, radio-iodinated (MIBG) 260
Misconduct 10–11
  see also Malpractice claims; Negligence
Mistakes, notification of 146
Monitoring departmental performance 134
  involvement of professional in 134–136
  risk management 143
Monoclonal antibodies 260
  F(ab) fragments 262–263
  labelled 266

Monte Carlo dosimetry techniques 28, 32
Moral dilemmas
  MRI, medico-legal work 155
  radiographic research 83–94
  ultrasound 58–67
Mortality
  common causes of death 21
  reduction, breast cancer screening 14, 15, 16, 17, 19–20
Multileaf Collimator (MLC) system 38, 39, 40, 41, 42, 43, 44
Myocardium
  imaging see Cardiac imaging
  saturation of 229

Named-radiographer system 100, 101
National Radiological Protection Board (NRPB) 18, 30, 31, 171–172
  annual occupational effective dose 169–170
  incidence of cancer 167, 168
  national reference doses 30–31
  patient protection recommendations 32
National Registry for Radiation Workers (NRRW) 167–168
Navigator echoes 230
Negligence 148, 149, 150–151, 156–160
  see also Malpractice claims; Misconduct
Network
  evaluating use of 43–44
  Philips RTNet 40, 41
NHS breast screening programme (NHS BSP) 14, 15, 19, 20, 22, 24
NHS and Community Care Act, 1990: 279, 280
Niche magnets 50
Non-maleficence and beneficence 59, 60, 61, 87, 90
Nuclear medicine, current trends 259–268
  advances in specific systems 266–267
  molecular and cell biology 259–262
  physics 263–266
  radiochemistry 262–263
  see also Gamma camera imaging, positron emitting isotopes
'Number needed to treat' (NNT) concept 16–17

Observation in research 88, 90
Outreach clinical services 123–129
  development of 124–126
  facilities 126–127
  the future 128–129
  objectives 127–128

PACS (Picture Archiving and Communications System) 253, 290
Palliative radiotherapy 108–109
Pathology terms, classifying in report 74
Patient's Charter 96–103, 116, 117, 131, 134, 279
  and explanations, care and treatment 142
  policy agenda 98–99
  professional agenda 100–101
  publicity agenda 99–100
  radiography services 101–103
'Pattern recogniser for iris of exposure field' (PRIEF) algorithm 291, 293
Pay and conditions of employment 3, 4
Periventricular leukomalacia 151–152
Personal action plan/career plan 275
Personal injury firms 149
PET scanners 37, 174, 175, 177–178, 183, 190, 191
  attenuation correction 178
  breast tumours 199
  count rate capability 182
  detection sensitivity 180, 181
  FDG imaging 187, 188, 189, 191
  recent developments in technology 185–186, 265–266, 267
  schematic view 176
Phase wrap artefacts, MRI 212
Phased array radiofrequency (rf) coils 49–50
Philips Multileaf Collimator (MLC) 38, 39, 40, 41, 42, 43, 44
Philips Stereotactic Collimation system 38, 39
Planar gamma camera FDG imaging 186–187
Planning see Treatment planning
Portable ultrasound scanners 244, 246, 247, 254
'Pre-read' scan, computed radiography 291, 292

Pregnancy
  radiation dose limits 166, 169, 171
  termination of 63, 64–65
Press, breast cancer screening 15, 24
Professional
  development and service needs 270–277
  changes in workplace environment 271
  framework for development 272–275
  value to organization 276–277
  see also Education; Training
  influences in contracting 130–137
Professionalism, ethics and research 83–84
Professions
  characteristics of 4–6
  conditions in common 6–8
  definition of 6
  obligations to the public 8–11
  positive and negative perceptions 283–284
  status 284
  see also Status of radiography
Professions Supplementary to Medicine Act, 1960: 4, 7, 8, 11
Prostate cancer, dose escalation study 42
Proximal coronary arteries, luminal diameter 231, 232
Psychotherapy, radiotherapy patients 108, 109, 118

Qualifiers in image reports 73–74
Quality of care
  contracting for 130–137
  improved, shared learning 279–287
  The Patient's Charter 96–103
Questions, research 90

Radiation dose limits 164–172
  development of 164–166
  dose limits since DS 86: 168–171
  evidence used in determining 167–168
  need for 164
Radiation Protection Adviser (RPA) 126, 127
Radiography News, definition of profession 6
Radiological protection, CT 26–33

Radiotherapy
  breast 215
  patients' perceptions of treatment 105–112
  prostate 42
  technology and changing practice 35–46
Random trials
  controlled 88–89
  large number 16
'Red dot' systems 79
Reference doses, CT 30–31, 32
Remote screening centres 257
Report of the Health Service Ombudsman 117–118
Reporting *see* Image reporting
Research
  ethical dilemmas in radiographic 83–94
  and evaluation, new technologies 43
  the Macmillan radiographer 118–119
  Society of Radiographers 7
Respect, principle of 86
Rib rotation
  measuring 242, 243
  pre- and postoperative 245
Rights and Standards, Patients' Charter 96, 97, 98, 100, 101, 102
Risk management 139–146
  analysis of risk 140–141
  control of risk 141
  definition 139–140
  in health care 141–144
  identification of risk 140
  in radiography 144–146
Role of
  nurse 100
  radiographer 35–36, 137, 145, 281
Royal College of Radiologists 75
  image reporting 81
  skill mix, guidance on 271

Scattered radiation
  computed radiography 295
  detected counts, PET scanner/gamma camera 176, 184
School screening for scoliosis, ultrasound 238, 241
  real-time portable machine 246
  rib/spinal rotation 245
Scintillation detection (PET) 178
Scoliometer 238, 240

Scoliosis 237–238
  role of ultrasound 239–247
Sedation, MRI of children 155
Segmented k-space MRI technique 229
Service level agreement (SLA) 132
Shared learning 279–287
  the context 280–281
  future development of profession and ISPL 285–286
  identity, boundaries and ISPL 283–285
  interprofessional shared learning 281–283
  in undergraduate education 286–287
Side-effects, radiotherapy treatment 107–108
Signal to noise ratio (SNR)
  computed radiography 295
  MRI 49, 50, 53, 56
Signal void, coronary artery abnormality 232
Silicone breast implants, MRI 218, 220
Simulators 37
Skin, dose equivalent limits 168–169, 170
Society of Radiographers 1, 2, 5, 10, 21, 136
  members of Council 2, 3
  misconduct 10, 11
  research 7, 93
Sodium iodide (NaI) scintillator 178, 186
'Soft markers', fetal ultrasound 61–62, 67
SPECT (single photon emission computed tomography) 175, 176, 178, 181, 187, 188, 189, 263–264, 265
Spectroscopy 18, 56
Spinal rotation
  measuring 242, 243
  pre- and postoperative 245
Spine, under load, MRI 51
Stages of research programme 86–91
  data collection 89–90
  data processing 90–91
  participant selection and assignment 88–89
  publishing results 91
  research area and methodology 87–88
Standards
  quality of care 133, 135
    multidisciplinary 134
  risk management 144
  teleradiology 253–254, 255

Standards (*Continued*)
  see also Rights and Standards, Patients' Charter
Status of radiography 1–12, 93
  dichotomy of interests 3–4
  medical dominance 2–3
  re-establishment as profession 4
Stereotactic radiotherapy 38, 39, 40
Stochastic effects of radiation 165
Superimposition of gamma camera/PET images 264–265
Supportive care, Beatson Oncology Centre 119, 120
Surface coils, MRI of breast 207
Surgery
  breast cancer 197, 212, 215, 225
  image guided (MRI) 51, 52
  minimal access 252
  scoliosis 246

Technetium-99m 179, 186, 187, 259–260, 264, 267
Techniques
  MRI
    brain 153–154
    breast 205–212
    coronary artery 229–230
  ultrasound
    back shape 242–244
    femoral anteversion and tibial torsion 244
Technology and changing practice, radiotherapy 35–46
  efficiency 43–44
  future departments 36–40
  new technologies 45
  radiographer's role 35–36
  skills required 44
  training issues 44–45
  treatment techniques 40, 41, 42–43
Telemedicine 250
Teleradiology in primary care 250–257
  advantages and disavantages 255–256
  need for 251–252
  needs met so far 252–254
Termination of pregnancy 63, 64–65
Thallium perfusion studies 187
Thermography, breast 199
Three-dimensional (3D) computer planning systems 37, 42

Three-headed gamma cameras 188
Tibial torsion, ultrasound evaluation 241, 244, 246
Time signal intensity curve, MRI breast enhancement 201–202, 203, 204, 222
Tracers, nuclear medicine 259–260
  FDG (fluorine-18 fluorodeoxyglucose) 174, 179, 181, 186–189, 190, 191–192, 266
Training 137, 277
  image reporting 79, 80
  new technology 44–45, 145
  psychosocial care, cancer patients 110, 117–118
    counselling skills 118
  see also Education; Professional: development and service needs
Transillumination, breast 199
Transmission
  computed tomography 178, 184
  imaging, PET scanner 185
Treatment planning 36–38, 40, 42, 45
Trials see Random trials
Tumour markers, C-11 amino acid imaging 190–191

Ultrasound
  breast 199
  fetal anomalies, moral dilemmas in diagnosing 58–67
    18–20 week routine scan 59–60, 63
    comment 66–67
    communication of results 63–64
    conscientious objectors 64–65
    informed consent 60–61
    morals and moral dilemmas 58–59
    patient versus society 63
    'soft markers' 61–62, 67
  outreach services 128
    equipment 124, 127
    image reporting 127
  scoliosis, role in diagnosis and management 237–247
    discussion 246–247
    groups of patients 239
    methods 242–244
    previous work on posterior trunk 239
    radiological methods of investigating 240

Ultrasound *(Continued)*
  scoliosis, role in diagnosis and
    management *(Continued)*
      summary of findings 245–246
      surface methods of investigating
        239–240
      ultrasound applications 241
      use in general practices 254

Verification, treatment 37–38, 45
  on line 42–43
Video presentation, informing patients
  108

Waiting times 98
  imaging results 251
  radiotherapy departments 106
Welsh Health Planning Forum 254,
  271
  benefits of teleradiology 251–252
  creation of long distance links 253
Whole body dose limit 165, 166, 170

X-ray examinations, patient dose 27
  see also Computed tomography (CT):
    radiological protection
X-Ray Quality Circle 102